A TOUR OF MONT BLANC

A TOUR OF MONT BLANC

Icons by Rupert Taverner

Summersdale Publishers Ltd
46 West Street
Chichester
West Sussex
PO19 1RP
UK

www.summersdale.com

Printed and bound by CPI Group (UK) Ltd, Croydon, CR0 4YY

ISBN: 978-1-84953-519-9

A TOUR OF MONT BLANC

AND OTHER CIRCUITOUS ADVENTURES IN ITALY, FRANCE AND SWITZERLAND

DAVID LE VAY

summersdale

For Nicky and Jessica
And for my parents:
Abraham David and Sonja Johanne

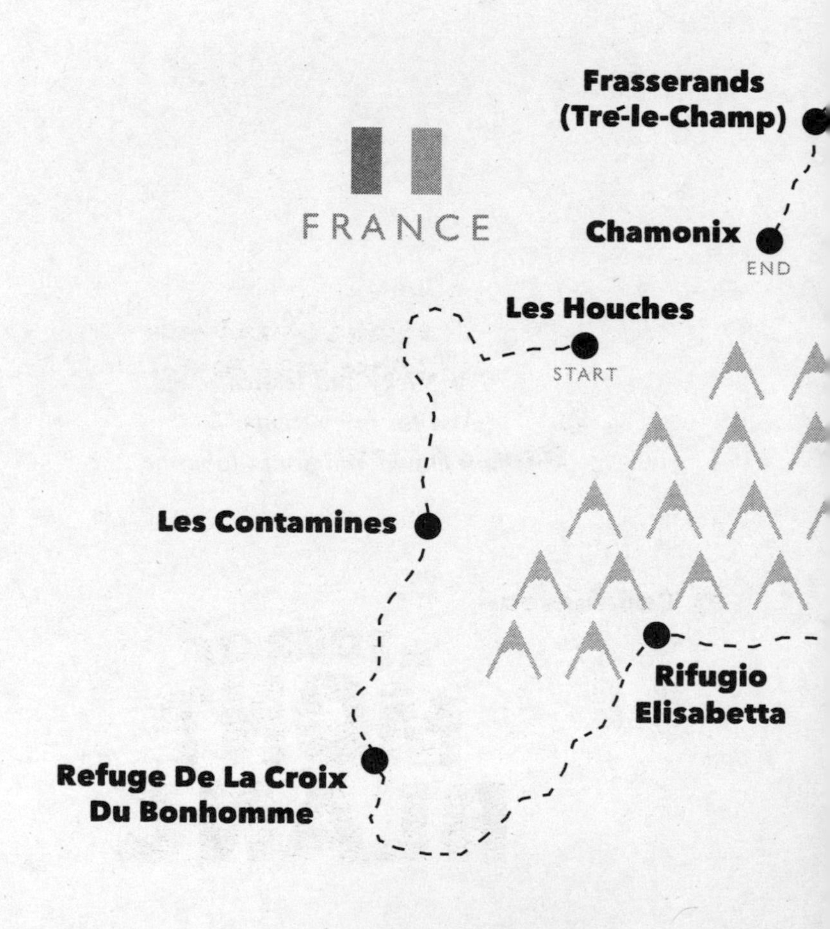

Frasserands
(Tré-le-Champ)
FRANCE
Chamonix
END
Les Houches
START
Les Contamines
Rifugio
Elisabetta
Refuge De La Croix
Du Bonhomme

Col De La Forclaz

Champex

SWITZERLAND

Alpage De La Peule
(close to La Fouly)

Rifugio Bonatti

Courmayeur

A TOUR OF MONT BLANC

Contents

Prologue

Mont Blanc, or Il Bianco as it is sometimes known in Italy; the White One sits regally upon pleated glacial folds gathered together around a timeless, grey-rock body, high above the French town of Chamonix; queen of all that it surveys. It beguiles, bewitches and charms the lesser folk who seek to frolic and play among the buttressed foothills of its ancient frame, exuding also a sense of foreboding; a warning to the one hundred or so climbers that die each year attempting to reach the gleaming, domed summit of this great snow-encrusted mountain.

Framed by the cobalt blue sky and golden alpine sun the mountain is a benign presence, reassuring in the sheer scale and might of its ageless form. But like any great mountain, its mood fluctuates with the attendant courtiers of rain, wind, snow and ice that make it capable of unpredictable flashes of malevolence that demand an experienced vigilance and vigour from all those who fall under its wintry gaze. At 4,810 metres, Mont Blanc stands proud and tall but not alone as it casts a wary eye over rival siblings that jostle impatiently for attention. The Grandes Jurasses, Aiguille Noire de Peuterey, Aiguille Verte, Les Drus, Aiguille du Midi, Mont Maudit and Mont Dolent all form part of this great mountain range, a stunning 25-kilometre wall of rock with its 400 individual summits and over 40 glaciers which, like crystalline jewels, bedeck the frontiers of France, Italy and Switzerland. There is a power in these mountains and, as I crane my neck and shield my eyes from the glare of the morning sun, one cannot help but be captivated, gripped and awestruck by the sheer presence and scale of these great prehistoric giants.

Mountains do something to me; they can transform one's perspective and their external grandeur is mirrored with something more internal, a contemplation of one's own place in this world that somehow lays waste to the petty troubles, worries and anxieties that we carry with us like some kind of emotional, modern day cilice. It is as if when I am not walking in these places I carry another kind of rucksack, invisible yet twice as heavy as the 15 kilogram one that I now carry on my back and it is with a joyous relief that I can snap the metaphorical buckles of this burgeoning load and let it drift away into the autumn mountain ether.

There is also a dark psychology about mountains, something to do with their primitive timelessness and the bipolar nature of their shifting temperament; elemental forces that carry echoes of the unconscious, the Jungian shadow perhaps. The Romantic poet, Percy Bysshe Shelley wrote of Mont Blanc in 1816: 'the immensity of these summits excited when they suddenly burst upon the sight, a sentiment of ecstatic wonder, not unallied to madness'. It was as if Shelley recognised something of himself reflected in the visceral glare of the mountain's intensity. In his poem 'Mont Blanc: Lines Written in the Vale of Chamouni', Shelley writes:

> *Power dwells apart in its tranquillity,*
> *Remote, serene, and inaccessible:*
> *And* this, *the naked countenance of earth,*
> *On which I gaze, even these primeval mountains*
> *Teach the adverting mind.*

This was a prevalent Romantic theme of the period, the inextricable connection between mind and nature and the notion that there is no purely objective reality, but it holds a truth for me now as I gaze upwards at the Mont Blanc massif and absorb its icy grandeur. Interestingly, Shelley was the husband of Mary Shelley, she of the gothic novel *Frankenstein*, which itself is an existential account of light and shadow and the nature of transformation. It is interesting how mountains, like Mont Blanc, can through their sheer infinite presence carve patterns into our understanding of the world around us and ourselves, shaping us in the same way that the mountains themselves have been shaped by the prevailing elements.

What I am saying here is that to experience the mountains is to experience a part of oneself. It is an encounter and like all encounters leaves one changed in some small way (or perhaps not so small). And, of course, long distance walking itself is a form of meditation, each step in itself part of the seductive mantra of motion and movement. So this book is about both walking and mountains as I embark on the famed Tour du Mont Blanc, a 170-kilometre-long distance hike that circumnavigates the White Lady of Mont Blanc. It is a metaphysical journey as much as it is a purely physical one as, like all journeys, it is a process of contemplation and self-reflection. So forgive me my narrative whims and my circuitous, tangential forays down life's many country lanes. Sometimes one doesn't truly know where these journeys begin and end, living as we do somewhere in the middle.

One

A little like nesting turtles, at the turn of every year my family and assorted friends return to a magical part of Cornwall that nestles reassuringly among the nooks, crannies and creeks of the Helford estuary and hatch out new, ill-advised plans, buoyed by sparkling wine and the ice cold water that we ritually plunge ourselves into soon after midnight in a bid to cleanse ourselves of the previous year's hangover.

It's a cold, wet but deliciously fresh New Year's morning as my good friend Rupert and I decide to hit the coastal path for the brisk two-hour walk to Falmouth. The path is muddy and well trodden from the assorted walkers who, like us, are gratefully allowing the invigorating, brine-infused sea air to blow away the few remaining cobwebs from last night's end of year festivities. As we cautiously slip, slide and squelch our way across the sodden fields, we reflect upon our half-remembered late night conversation about new potential challenges for the year ahead. It seems to be the case that most of my madcap adventures begin their life on a tiny, secluded Cornish beach in the early hours of the New Year. My last big venture, an

850-kilometre coast-to-coast trek along the French Pyrenees, was brought in to being over a pint or two in the local here, and my partner Nicky has to think twice now about these Cornish trips as she never quite knows where they might lead.

I learned a lot during my seven weeks on the GR10 (Le sentier de grande randonnée dix) in the Pyrenees. Thinking back there certainly was more than a whiff of midlife crisis about that walk, when I was in my mid-forties and frozen in the halogen daze of life's rich glare. Yes, I liked the idea of a coast-to-coast walk that bridged the Mediterranean and the Atlantic, two great oceans stapled together by the vast reach of the Pyrenees mountain range; I wonder on reflection what it was I might have been running from. It was a linear walk, point to point, and so by definition meant it was taking me away from one place as much as it was leading me to another and perhaps as I scaled the perilous slopes of my mid-forties this was what I needed: a sense of escape.

So as we walk the ever-scenic, timeless South West Coast Path towards the grey cranes of Falmouth harbour that float eerily in the early morning drizzle, I tentatively propose to Rupert the idea of walking the Tour of Mont Blanc (TMB) later in the year. What attracts me now about the TMB is its very circularity, the idea of circumnavigating the highest mountain in Western Europe. I want to embrace this great white mountain, encircle it with my arms and clasp my fingers tight around its icy belly. Hell, I might even rip it right out of the bloody ground and toss it over my left shoulder for good luck.

The days leading up to this walk around Mont Blanc have in many ways been very significant and perhaps a little portentous. Firstly, I reached the precarious age of 49, the stage at which we haul ourselves kicking and screaming up onto life's snow-capped summit and take our first nervous glimpse at what lies beyond. (If this is indeed the summit?) As we walkers well know, just when you think you have reached the peak of your climb you see a further one rising in the hazy mists of the far distance. But it feels strange to be approaching 50 and the majority of comments from friends and colleagues are generally along the lines of 'you're nearly there', 'almost hit the big one' or 'soon be a member of the club'. Well, perhaps I will one day gratefully pull on my old man's slippers, suck on my care-worn pipe and settle back into my far too comfortable armchair when I am a fully qualified member of 'the club' next year, but at this point in time it feels like a difficult final scramble over the overhang of the late forties to this particular peak on life's path. And one can't but help take a backward glance over the shoulder towards the flatlands of one's childhood and the rising foothills of adolescence. Perhaps more an instance of don't look back rather than don't look down.

Secondly, just a couple of weeks after I return from walking the TMB, our daughter Jess packs her bags and heads off to university to embark on a whole new stage in her young life. It is the beginning of her own long distance hike and she even has the half-packed rucksack sitting in the middle of her bedroom floor to really bring the analogy home. Perhaps I should give her my water bottles and Day-Glo orange survival bag (they might come in handy for fresher's week). But it is another defining moment; a punctuation of sorts wherein both mine and Nicky's immense pride and joy at seeing her embark on this exciting passage in her life is mixed with a twinge of apprehension as we pause and reflect upon its meaning. It is an inevitable phase of the family life cycle, a transition that requires a degree of negotiation, or perhaps renegotiation. Being our only child this moment carries more significance, and while Nicky is very open about her own process and the task of saying goodbye (or *à bientôt* as we would prefer to think) I wonder about the denial of my own feelings and whether there is, as I think about it now, a connection between Jess leaving home and my going on a long distance hike at this point in time.

I jokingly once said to a friend that my way of coping with moments like this was to run away, or at least I thought I was joking. All men live in a state of 'quiet desperation', to paraphrase the writer Henry David Thoreau, and as a man and a father I guess I must seek my own ways to manage my own occasional feelings of disquiet, ways to regain a sense of emotional balance and equilibrium. And walking in the mountains is one way I have found of achieving this because I find that the process of engaging with physical perspective transforms my own sense of internal perspective. It is also the act of movement and momentum; just as a young child, when struggling to convey an idea or concept will jump, wriggle and pace the floor as thought and action, mind and body, become entwined together in a wonderful dance of attunement. As Thoreau wrote in one of his journals:

> *As a single footstep will not make a path on the earth, so a single thought will not make a pathway in the mind. To make a deep physical path, we walk again and again. To make a deep mental path, we must think over and over the kind of thoughts we wish to dominate our lives.*

So, standing in a muddy field in the New Year's morning breeze, with the blue-grey waters of the Helford estuary opening up ahead of us, Rupert and I shake hands on our agreement to take on the Tour du Mont Blanc. These are the important deals that one makes in life; not the fickle Faustian pacts of modern life that slowly demean and diminish our soul, which we sell in piecemeal fashion to the corporate devil. No, this is a simple deal of friendship and adventure. It is a deal that brings a glint to the eye, childlike in its innocence, as the two of us benignly collude in plotting a temporary escape from the relentless pressure of work and responsibility. This talk of Faust may perhaps sound a little dramatic, but I do believe that we all lose a little bit of ourselves in the never-ending chase to keep up with the travails of our contemporary lives. We lose time, sleep, the quality of our relationships. We lose our hair, our fingernails, our health, sometimes our friendships. It is, as we are so often told, a price we have to pay for maintaining the 'good life' though I wonder sometimes whether the gains of this good life are worth the cost.

Ultimately, it is about meaning; whether we feel fulfilled, enriched and sustained by what we do and how we do it. Of course there are gains; my work as a child therapist is certainly both meaningful and enriching, but it is also stressful and often emotionally depleting and so I seek ways in which to maintain some kind of healthy balance and outlook upon life. So the idea of these long distance hikes is not so much about finding oneself, more about regaining something that might have been lost or mislaid along the way, a sense perhaps of value and perspective.

The thought of walking with Rupert is also good; I count him as one of my closest friends and he will be a good travelling companion. We share a lot in terms of our values, interests and general outlook on life and in the midst of the ups and downs, both literal and metaphoric, that are an inevitable part of any long distance trek, I think we will be able to encourage and spur each other on towards the finish line. We are also at ease in each other's company; able to share a laugh and not take ourselves too seriously while also acknowledging that life can be… well… serious. We both work at the sharp end of human experience and in this sense relish the opportunity to lay down our tools, breathe in some mountain air and reconnect with the world around us.

Two

The Tour du Mont Blanc has established itself as one of the world's classic, long-distance mountain walks and is certainly the most popular walk of this kind in Europe, enticing more than 10,000 eager trekkers each year to embark on this 170-kilometre circuit around the magnificent Mont Blanc massif. The fact that it can be completed in anything between eight to twelve days also makes it a very attractive option for those looking for a challenging walk that can be completed within a reasonable time. Certainly, the idea of nipping around Mont Blanc in a week or so goes down a little better at home than taking off for two months in the Pyrenees – in my experience anyway.

The first recorded tour of Mont Blanc was actually undertaken way back in 1767 by Horace-Bénédict de Saussure, the geologist, botanist, physicist and general all-round Alpine explorer and aristocrat. Saussure first visited Chamonix in 1760 and was so impressed by Mont Blanc that he offered a reward of twenty thalers (the currency of the time) to the first person to conquer the great mountain. It took a further 26 years for his reward to be finally claimed when two Chamonix men, Michel-Gabriel Paccard and Jacques Balmat, finally reached the summit in 1786. Saussure himself almost laid claim to his own reward, attaining the summit just a year later in 1787. He was that rare breed of polymath, a man of the era I suppose, whose intellect and unquenchable curiosity spanned all manner of disciplines, unlike today where specialisms seem to be quite the thing. I wonder sometimes if we are missing a trick by having an education system that corrals people's natural intellect into such narrow and specific areas, but perhaps that is a discussion for another day.

Saussure, for all his intellect, has also been described as the 'inventor of climbing' and the founder of alpinism, again harking back to the golden age of exploration when both science and adventure were combined in equal part. Apparently though, he didn't travel lightly and a list of Saussure's 'items to be taken to Mont Blanc' included two frock coats, several waistcoats, two nightshirts, three pairs of shoes, a pair of slippers and two cravats. And in addition to a bed and blanket he also took a mattress and curtain. Funnily enough, this is not dissimilar to Rupert's own travelling wardrobe.

Anyway, 60 years later Saussure was succeeded by J. D. Forbes, Professor of Natural Philosophy at Edinburgh University, who made his own tour of Mont Blanc in 1839. Forbes was something of a glaciologist and clearly combined his thirst for knowledge and discovery with his love of mountains and was particularly interested in the phenomena of thermal energy and glacial movement and made quite a name for himself in both these areas. Indeed, the Aiguille Forbes high above Chamonix was named after him, as were the Forbes Glaciers in New Zealand. The TMB became increasingly popular during the late 1800s, when the transport of choice was more often than not a mule, and then in later years established itself as one of the world's top mountain walking trails, particularly as the activity of trekking and backpacking rapidly grew in popularity.

Talking of new trends, there is another tour beyond the traditional TMB; the UTMB, the popular ultra marathon that takes place every year in which competitors run the complete circuit around Mont Blanc, day and night, with the fastest runners completing the race in just a little over 20 hours. Now, I like a bit of cross-country running and compete in the occasional 10-kilometre cross-country and half marathon myself, but if you ask me this ultra marathon is a race for nutters. In fact I am deeply suspicious of anything, especially sport-related, that has the prefix 'ultra' attached to it, unless of course it's table tennis... but then I am not quite sure how ultra table tennis would work.

Mountains do tend to attract the extreme end of the sporting spectrum and there is another more recently established race in this region called the Tor des Géants, a 330-kilometre endurance trail that starts in Courmayeur, Italy. This tortuous trail takes you around the four great giants of the Alps: Mont Blanc, the Gran Paradiso, Monte Rosa and the Matterhorn with a total height gain of 24,000 metres, which the competitors complete in around 150 hours, and makes the United Kingdom's Three Peaks Challenge look like a walk in the park. Nutters the lot of them.

So call us boring, but Rupert and I decide to take on the TMB in a more conventional manner, namely walking, and leave the more extreme, ultra stuff to the bright young things. Actually, we aim to begin the walk on the first weekend of September, the very same weekend that the ultra marathon ends, our initial research suggesting that it is generally best to avoid the UTMB due to the demands upon accommodation over the period of the race itself. But we might possibly brush shoulders (or rucksacks) with a few of the runners or even follow in their eager slipstream.

The TMB can be walked clockwise or anticlockwise, the latter being the favoured mode. There does not seem to be a very persuasive or specific reason for this other than the fact that the traditional starting point for the walk is the small town of Les Houches in the Chamonix valley, about seven kilometres southwest from Chamonix itself. To walk clockwise from Les Houches would mean having to take on the steep 1,500-metre climb up to Le Brévent, which would be a tough first day for any walker, so better to head off in the other direction. Walking clockwise, against the flow as it were, also means that you would periodically meet waves of walkers coming from the opposite direction and more than anything else this would involve a lot of hellos, *bonjours* or *ciaos* and I don't think I could cope with that for two hours each morning. It would also follow that you would meet a new group of walkers at each night's rest point. But walking with the flow, anticlockwise, you would be meeting many of the same people following the same stages as yourself and my experience when walking the GR10 was that hooking up with fellow walkers along the way was one of the great pleasures of the journey.

The definitive guidebook for this walk is *Tour of Mont Blanc: Complete Two-way Trekking Guide*[1] by Kev Reynolds and it is such a good book that we will do more or less anything that Kev suggests. He sets out a good, clear and manageable stage-by-stage route plan along with plenty of sound, informative advice. So following on from our conventional decision to walk rather than run the TMB, we continue in our conformist ways and decide to take the anticlockwise option.

So we know when we are going to go and which way round we are going to go, the next decision being how we are actually going to get down to Les Houches itself, which sits about 75 kilometres southeast of Geneva. Wary of our carbon footprint we dismiss flying, the cheapest and simplest option. The train, our first choice, is too expensive so we decide to drive down via the cross channel ferry at Dover. Like any meaningful journey, the process of travelling is as important (if not more so) than the actual destination itself. This may sound trite but it is something easily lost within the often frenetic routines of our daily lives. The way I see it, the challenge in life is to stay in the moment, to slow down and open oneself up to the experience of what is happing in the immediate present, rather than becoming over preoccupied with the past or future. I know myself that I can too easily become restless, agitated and angst-ridden to the point sometimes when I cease to function and get hit by treacly waves of gloomy inertia. It is as if the thought of all the things I should or could be doing actually result in me doing nothing at all, like some kind of existential paradox. It is not depression, but perhaps not too far from it either.

It was probably not until my early twenties that I truly understood the fact that my father suffered from quite severe mental health problems, his mood swinging between deep depression and a fizzing mania, but as a young child I grew up in a house where the thick, dark, suffocating blanket of his depression was never that far away. Interestingly, one of my father's many ways of self-medicating was through walking and while it is a reflective activity it also slows you down and grounds you in the present. You become aware of the immediate reality of your environment, notice the fine detail of all that surrounds you as well as becoming acutely attuned to the physicality of your own body. It is, as we therapists say, all about process. Where you want to get to is clearly important, but how you get there possibly more so. Where Rupert and I want to get to is Les Houches and having dismissed train and plane it is going to be car and ferry, having also quickly dismissed the channel tunnel. You can't beat a ferry for getting a sense of travelling somewhere, even if it is just the short hop across the water to Calais.

Overall, our preparation for this walk has been left to the last minute, the timing being dependent upon Jess' exam results in mid-August, which determine which university she will be heading off to and, most importantly, when. We aim to go at the beginning of September to avoid the ultra marathon and also the school holiday period, when things might be a little quieter on the trail. Jess has an offer of a place at Cambridge, where she would start at the beginning of October. However, if she doesn't get her grades then she will be off to Manchester in mid-September, which would mean postponing our walk for another time – most likely next year as I can't imagine not being around when Jess leaves home, alongside all the practicalities of getting her and her various belongings safely relocated to university. So the plans for our walk have become inexorably linked with Jess's exam results. No pressure I tell her, but you had better get those grades. She just gives me a withering look, mumbles something about priorities and suggests what I might do with my water bottle. As it is, on the day after my forty-ninth birthday, she gets the required results so roll on Cambridge and the TMB.

This gives us just a couple of weeks to get our kit together, although in relation to the key items such as boots, rucksack, sleeping bag, etcetera, I am all but sorted as I still have these from my earlier trek on the GR10. Rupert though has to start pretty much from scratch, not least with his walking boots, which have become the source of much consternation. Now, tales of Rupert's shoe buying escapades are legendary and I kid you not when I say that there is even a distant story about him being chased out of one particular footwear outlet by a furious man with a shotgun, so enraged was he with Rupert's shoe-based prevarication. So when it comes to buying a pair of walking boots Rupert has form, and you could well imagine shoe shops up and down the land sticking posters up in their window: Beware this man; serve with caution. The build up to our departure and gradual acquisition of our respective kit was rather dominated by talk of Rupert's walking boots and how many he had tried on and taken back in the search for 'the one' (or perhaps I should say 'the two').

Our journey starts in a lay-by next to Wisley Gardens on the Surrey stretch of the A3 at 5.30 a.m. where I have arranged to rendezvous with Rupert who has kindly volunteered the use of his car for the trip. Nicky offers to give me a lift to our meeting point and it is with a strange mixture of excitement, apprehension and tiredness that we wait together for Rupert.

The thin light of the cold morning slowly coalesces around us and heavy traffic, even at this time of day, hurtles past and then after a short while we see Rupert pull in and park up ahead. I allow the tiny part of me that momentarily imagined that he might not turn up so that I could go back to my warm, Saturday morning bed, to drift away into the early morning mist. This looks like it then.

Three

After a fond farewell to Nicky (she is getting used to them now), Rupert and I head off towards Dover, and although we are late we feel relaxed about it. I tell Rupert about a gangster book I once read in which the central protagonist, an Italian Mafioso-type figure, lived by the expression 'the thing is the thing'. I have quite taken to the phrase, in the sense of having to accept the reality of one's situation instead of forever hankering after something better or different. I have even woven it into lectures to my students at university when discussing the relative merits of analytic interpretation within children's symbolic play, but they just stare at me blankly and wonder if I have finally lost the plot. The thing is the thing? Anyway, Rupert and I decide to adopt a 'the thing is the thing' philosophy for our trip, which means the concept of lateness suddenly ceases to exist. We will arrive when we arrive. As it is, when we do arrive we are told that a ferry has broken down, the timetable is in chaos and they will try and get us on the next ferry that has space, however long that may take. Some may have been disheartened by this situation but Rupert and I have arrived with the perfect mind-set... the thing is the thing.

So after a couple of hours of drinking coffee and idle chat we board the next available ferry and say farewell to Old Blighty and the gradually diminishing White Cliffs of Dover. There is something about ferries that always induces a slight frisson of childhood excitement; the sound of the engines churning away deep below, the oily smell of diesel, the sights and sounds of the crew casting off with chains and ropes the size of tree trunks and finally that moment when you look over the deck railings and see the water zipping past far below. With Nicky and Jess I have spent many a time travelling around the Greek islands on ferries, which is always a delight, and I also remember when as a young child, perhaps six or seven, we went on a family holiday to Spain by ferry and got caught in a great storm in the Bay of Biscay. Practically everyone was struck down with seasickness and hid away in their cabins while I seemed somehow immune and roved around as if I had the whole boat to myself. It's a good way to travel and, in the humble view of Rupert and myself, so much better than taking the tunnel.

We are not sure how far we might get today; possibly all the way to Les Houches although at this time of day it's unlikely, so we decide to just see where we are by around 8.00 p.m. and then call it a day and find somewhere to stay. We are disgorged unceremoniously at Calais and strike out on the A25, southeast towards Reims. As always, it feels great to be in France again. We make good progress, taking turns to drive and stopping off for a bit of late breakfast/lunch on the way. We both have a strong desire for a croque-monsieur, that most iconic of French snacks, and somehow just saying 'croque-monsieur' feels reassuring and evokes everything that is good about France. The urban myth surrounding the croque-monsieur is that it came about by the somewhat fortunate accident of French workmen leaving their ham and cheese sandwiches next to a hot radiator all morning and then rather enjoying the gooey result. True or not, the humble croque-monsieur deserves its place in French culinary history.

But after all that fanfare, the cafe doesn't have any so we have to satisfy ourselves with a simple ham and cheese baguette. Could we leave them next to the heater? Our ham and cheesy break over, we continue on our way and get as far as Dijon (which somehow seems rather fitting, it being the national home of French mustard) where we decide to call it a day and stop for the night. We are both tired and hungry and after driving around eventually find a cheapish place to stay. It is late by the time we walk into town and everywhere seems closed so we have to make do with a McDonald's and a couple of beers. From the sublime to the ridiculous.

We leave Dijon at around 8.00 a.m. and have a good run down the A39 to Les Houches. The French motorways are, as ever, a pleasure to drive on although it is a pleasure you have to pay for. On balance though, I think it is well worth the price as we sweep down the empty road that stretches out ahead of us as far as the eye can see. We branch off eastwards on the A40 towards Geneva and the landscape, which has gradually shifted from flat to gently undulating, opens up into something much more dramatic as we get our first glimpse of the Alps, pushing into the sky ahead of us like snow-capped castle turrets in some kind of distant fairy-tale land. The weather for this transition into the mountain world that we will inhabit for the next ten days or so is perfect, a gorgeous blue sky and golden sun that bodes well for our walk. Rupert and I feel enthused by our first sight of the mountains and buoyed by the fantastic weather. Our plan is to find somewhere in Les Houches where we can safely leave the car and begin walking by early afternoon. We don't really want to hang around and lose another day and with the weather being as it is we feel that the first stage of the walk, about four or five hours, should be manageable even with a late start.

As we draw closer the mountains rise ever upwards and we get our first real sense of the Mont Blanc massif itself, a great cathedral of rock and ice, piercing the cobalt sky. Are we really going to walk all the way around this great White Lady, this colossus of a mountain? It seems quite a challenge as we look at it now and we feel both excited and daunted by the prospect of the journey ahead of us.

We arrive in Les Houches around midday and feel pleased with our progress. After a little drive around town to orientate ourselves we come to rest in the car park by the Bellevue cable car and start the process of gathering ourselves for the walk to come, sorting out what to leave and what to take as we abandon the comparative security of the car and head off on foot. I am as ruthless as possible, dumping everything that seems unnecessary including my raincoat in favour of my trusty poncho that stood me in such good stead in the Pyrenees. I remember well the psychological warfare that took place between my rucksack and I on our long weeks on the GR10. It was a hard-fought battle for control from which I eventually emerged victorious (if not entirely unscathed) and I want to do the best to ensure that this time round I give myself as good an advantage as possible. Rupert, on the other hand, seems intent on taking everything that he doesn't need: a selection of novels, trainers, sketch pad and pastels and a nice range of lotions and potions. I wouldn't be surprised if he has a table lamp, a couple of bottles of Pinot Noir and possibly even a kitchen sink in his burgeoning rucksack. I anticipate that Rupert will soon become embroiled in his very own battle of the bulge over the days to come.

Les Houches is the traditional starting point for the TMB as well as a popular ski resort with an impressive range of ski runs and pistes that extend to altitudes of up to 1,900 metres. These ski runs are in frequent use for international events with one run, the Kandahar, being used annually for the Men's Downhill World Cup Ski Championships. It is an attractive town, with its wooden chalet-esque buildings lending a slightly Swiss quality to the place. It's fairly quiet during this relatively sleepy season between summer and winter, but within that quietness there is still a gentle buzz of activity as people wander around with rucksacks, crampons, ropes and ice picks, reflecting the range of activities that draw people to these mountain regions. This place is all about people who like to climb, ski, walk, cycle, run and generally fall over a lot.

Four

The French Alps cover about 35,000 square kilometres of the Western Alpine chain. Imagine a great arc stretching from Lake Geneva to the Mediterranean, 350 kilometres long with an average width of 100 kilometres. Sitting at 4,810 metres, Mont Blanc is the highest point in this chain. I am glad that we are going around this great mountain, rather than up it, but the compulsion to climb as well as circle its flanks is surprisingly strong, an urge that I have to suppress and file away in that worrisome and expanding bottom draw that lurks in the dark recesses of my mind.

It's odd, the instinctive desire to climb a mountain 'because it's there' as the adage goes. But it's more than that; like sitting watching the iridescent flames of a crackling fire or lying on the warm stones of a beach listening to the gentle ebb and flow of the tidal waters. These are primal desires, base urges, like some kind of hangover from times gone by, sensual memories associated with the inherent needs of warmth, food, companionship and safety. But these trace memories and instinctual feelings also provide us with a visceral, elemental connection to the world around us and I feel a particular sense of being at peace with myself – a sense of feeling real or alive – when walking and immersing myself in these places. As our romantic friend Saussure said, 'the soul is uplifted, the powers of intelligence seem to widen and in the midst of this majestic silence, one seems to hear the voice of nature and to become the confidante of her most secret workings'.[2]

Putting climbing to one side (for the moment), I am also very drawn to the idea of walking around Mont Blanc. I like the idea of circumnavigating this great giant of rock, snow and ice. I am curious about the kind of relationship I and the mountain will have over the next two weeks; an endless sense of wonder and awe, but also the pain of physical endurance, the intense challenge of having to walk 170 kilometres – either uphill or downhill. I like the idea of seeing the mountain move with my ever-changing perspective; the light and shade, the structure, the folds and ridges and the sense that it will always be there; the constant reassurance of its presence like some kind of counterpoint to life's many uncertainties – a companion of sorts.

The structural characteristics of the Alps are defined by the enclosed valleys and long glaciers, the pinnacles and aiguilles – the needles – sharp crystalline-like structures of rock pushing into the air like great cracked, fractured spires of rock and stone. They remind me a little of the ice-blue copper sulphate crystals I used to grow in jars of water as a young child. The Alps are relatively young although still many millions of years old, and this puts my own existential anguish about hitting 50 into some perspective. But as Rupert and I ready ourselves for the trail that will lead us deeper into the ancient folds of the mountains I am profoundly aware that this circuitous journey around Mont Blanc seems poignant in so many ways; symbolic of the circularity of life, from birth to death – ashes to ashes perhaps. As I approach the zenith of my own long climb through the decades I am increasingly aware of my own mortality, a concept that up until now has seemed so abstract as to be nothing more than a vague unformed shape gently bobbing on the distant horizon. But strangely, as I manage now to hold the Reaper's gaze with more surety than I ever thought capable, I feel less – not more – concerned about the inevitability of finally shuffling off this mortal coil. As the saying goes, the problem is not death but the incompleteness of one's life and that perhaps is another part of the subtext that compels me to undertake these ventures, the idea that 'Old Father Time' is indeed running out.

Talking of ashes and fathers and personal subtexts to this journey, I have in the depths of my rucksack a small plastic tube containing some of my own father's ashes, smuggled through customs like some kind of illicit powdered narcotic… or powdered neurotic in my father's case. After he died our family planted a tree in his memory in the grounds of the village church in his beloved East Sussex and in with the tree went a few handfuls of his rather voluminous ashes that came in a large brown plastic container, similar in style to the original Elizabeth Shaw sweet tubs full of mint imperials or chocolate limes. For some reason we kept some of the ashes just in case. Just in case of what I am not sure; might there be some unforeseen future circumstance when we suddenly stop and think, 'Crikey, thank God we still have some of Dad's old ashes left.' So for the last few years he has remained in our attic, sealed genie-like in his plastic tub, alongside Nicky's parents who seem to have also fallen victim to the 'just in case' syndrome.

My father was quite a character and would have been quite at home among the early alpinists, explorers, scientists and romantic socialites who descended upon Mont Blanc and the Alps in the 1700s. A brilliant orthopaedic surgeon, prodigious writer, academic, linguist, musician, translator, traveller and, on top of all this, he was something of a great walker. My mother used to call him the 'wandering Jew' and his almost pathological restlessness led him (or drove him) all over the world.

Funnily enough, when he was in his forties and a few years before I was born, my father actually lived in Switzerland for a year or so while working for the World Health Organisation. So he wasn't quite Dr Who but was a WHO doctor. He lived with my mother and my elder brothers in a small village called Soral in the western tip of Switzerland and wrote of his hikes in the 'fierce and wild' country in the mountains above Visp at the far end of Valais. He told the story of J. B. Leishman, the 'great translator', who fell to his death from the very same mountain paths that he walked upon, thankfully my father was not following his exact footsteps. My mother recalls that from their house in Soral, when the valley wasn't filled with mist, they could see Mont Blanc and as I stand here now it is strange to think that all those years ago my parents were also living within its austere presence. Like a keeper of stories, a holder of histories, time flows around this mountain.

It's a beautiful sunny afternoon; a clear blue sky adorned with white clouds that perch like little fluffy hats upon the snow-capped peaks of the surrounding mountains. The September sun is surprisingly hot so we smother ourselves with suncream, fill our water bottles and finally hoist our rucksacks onto our protesting backs. Rupert has been slightly disparaging about my choice of walking gear, specifically my wooden stick and my hat.

My walking stick, or 'staff' as Rupert rather quaintly calls it, is a trusty veteran of my trek along the GR10 and there is a rough line etched into the handle for each of the 53 days that this epic journey took, a little like the days of a prison sentence scratched into the cold stone of a cell wall. I am very partial to objects and artefacts and my stick is one of my most prized possessions, so deeply attached to it have I become.

My hat can best be described as a floral trilby, a rash purchase some years ago at a music festival to protect myself from the sun while listening to Bob Dylan's rough approximation of singing (and God was it rough). That's the problem with attending festivals in one's middle age; you become lulled through a mixture of nostalgia, warm beer and rose-tinted specs into imagining that you are younger than you really are and end up buying something that you think is cool but actually just makes you look like an old bloke who... well, thinks he's cool. Hence the floral trilby. In a strange way I have become much attached to my hat as well, and at home the family know it as my 'gigging hat' because I wear it occasionally when playing with the band. My father was also very partial to walking sticks and eccentric hats, and when I take up my staff and floral trilby I somehow feel emotionally closer to him and perhaps a little like him. My relationship with my father was complex and never easy, but in a funny kind of way these long distance hikes are something of a tribute to him; a recognition of a part of him that I deeply identify with and so for that Dad, I tip my floral hat to you.

As we make our way down the road to the start of the TMB path that leads out of town, we both start to feel a little daunted by the challenge of the walk ahead. The route is about the same distance as once around the M25 and has an accumulated height gain and loss of around 10,000 metres and includes at least 11 mountain passes that have to be crossed as we move from valley to valley. This is no small physical challenge and it is important to be both physically fit and mentally prepared for the task. Having walked the GR10 (which I once read has a total height gain and loss of 49,000 metres – equivalent to climbing Everest over five times) I have an idea of what to expect and actually am much more physically prepared for this walk than I was for the GR10 for which I did no training at all. Rupert has done his fair share of walking over the years, but nothing quite on this scale so he is perhaps rather more concerned about the prospect of what lies ahead.

There are two possible routes for the first stage from Les Houches to Les Contamines, our planned stop for the night. One is a gruelling 18-kilometre walk that entails a massive climb of 1,478 metres and a descent of 1,318 metres. This is tough by any standards and as this is our first day and it is already early afternoon it would be foolish in the extreme to take this option. The second option is a much gentler 16-kilometre route that involves a height gain of only 650 metres and should take us about five hours. We walk through the outskirts of Les Houches and then turn left up a small road that soon turns into a rough track that begins to rise up through the surrounding woodland of fir trees and passes a few old wooden Savoyard houses.

After the frantic last week of preparation for this walk and the journey down through France it feels great to be on our way, although as the track rises steeply I get flashbacks to the days of relentless trudging up endless mountains in the Pyrenees and I have to remind myself that this is all part of the process. Although only 650 metres the ascent feels really hard going, not helped by the stifling heat of the afternoon sun that increases greatly as we slowly move out of the cover of the trees and lose the respite afforded by their occasional shade. The track weaves and hairpins its way up towards Col de Voza and every 200 metres or so we stop for a few minutes to get our breath back and to wipe the salty sweat from our stinging eyes. Occasionally we hear a warning shout from somewhere above, and minutes later two or three people on mountain bikes come hurtling down the track. During our fleeting eye contact I can see them thinking, 'Why on earth are you walking up when you could be doing this?' The same thought occurs to me.

Rupert is still very ambivalent about his boots and after only an hour or so of walking, the dreaded but long anticipated moment arrives when he stops and takes off a boot and sock to investigate an area of soreness that is developing on his heel. After a few moments of intense contemplation he solemnly pronounces that he has a blister. 'I knew these boots weren't right,' he says, 'I should have taken them back.' I am tempted to remind Rupert about his previous shoe shop escapades and that he might have a better chance of alighting upon the Holy Grail than finding the 'right' boots, but that might not be too helpful. Anyway, he applies a preventative plaster and we continue on our way, albeit a little more gingerly.

After a little over two hours we get to Col de Voza, which sits at 1,653 metres. The view from here is gorgeous, the surrounding mountains are encrusted with gleaming, metallic snow and we can look back down into the Vallée de l'Arve, bordered by a row of aiguilles. As is so often the way with this kind of walking the intense effort of the last couple of hours instantly falls away and we soak up the sheer beauty of our situation and feel boosted by a sudden euphoria. This is, after all, what it is all about: the pleasure of the physical exertion and the total glory of the mountains.

The col is rather dominated by a large hotel, Le Village Vacances, and we realise (with a strange mixture of regret and triumph) that we could have simply got here by hopping on the cable car from Les Houches, but of course that would not have been in the spirit of the journey. We rest for a while and have a bite to eat on the lovely and rather incongruous little station platform from which you can hop on the Tramway Du Mont Blanc. Inaugurated in 1909, the tramway is one of the oldest mechanical uphill rides in the region and the mountain rack railway is actually the highest in France, running from Saint-Gervais-les-Bains all the way up to the station at Nid d'Aigle that sits at the base of the Bionnassay glacier at an altitude of 2,232 metres. Overall, the line runs for just under 12.5 kilometres and for those of you that are interested (and I know you are out there) it has a rail gauge of 1,000 millimetres and is a combined rack and adhesion railway that uses the Strub design to overcome a total height differential of 1,792 metres. So there you have it. The line is actually invaluable for climbers who use it for the main French ascent up to Mont Blanc that starts from the tramway's terminus at Nid d'Aigle.

Feeling rested and just a little conscious of the time we cross the tramway and head downwards on a rough track that leads through an area of woods and open meadows and then drops to the delightful little hamlet of Bionnassay, the little wooden chalets adorning the hillside like a scene from Johanna Spyri's *Heidi*. Apart from Rupert's developing blister issue, the walking is pleasant and the path runs at a welcome downward gradient as we pass through a number of other charming little hamlets en route including Les Hoches, where a carved wooden sign on the front of a farmhouse reads: *Ferme Natale de l'Astronome Alexis Bouvard 1767–1843. Decouvreur de Neptune*. This is a lovely surprise and only confirms the already impressive scientific pedigree of this region.

The path is easy going; a mixture of metalled roads, rocky paths and woodland tracks and the late afternoon sun bathes the landscape in a magical warm glow. It is hard to absorb the beauty of our surroundings, the rich verdant meadows stippled with impressionistic brushstrokes of purple flowers and the darker green of the pine trees that point the way up to the dramatic grey rock of the mountains with their covering of snow that dazzles the eye in the bright light of the afternoon sun.

Being September, we are unfortunately not party to the full range of flora that adorns this region. The regional and often very local variations in soil, climate and altitude contribute to a rich diversity of plants and flowers, the mountains and valleys often creating their very own microclimates. In spring, crocuses and anemones sprinkle the green of the lower valleys with a dusting of blue, white and purple and higher up you may see the pink-blue of the tiny alpine snowbell. In summer, lilies and gentians are in abundance in a variety of forms and even now in early September pink-red alpenrose shrubs and creeping azaleas carpet the ground. And then there is the grey/white edelweiss, the national symbol of Switzerland. Snows allowing, it would quite something to do this walk in April or May and really experience the full range of these beautiful mountain flowers.

As we gently descend through woodland we follow the course of a lively river and as we cross over on a small wooden bridge we just have to stop for a while to soak up the pure, sensual beauty of the place and breathe in the rich air. It has been very quiet on the trail, I guess because of our late start, and the only other hikers we have seen are a charming young couple with whom we occasionally exchange a few words and a smile as we dovetail our way down the hillside. Eventually we reach the outskirts of Les Contamines and having been walking for close to six hours we are both beginning to feel exhausted. Rupert's blister has gone nuclear and is causing him considerable grief and so we are disheartened to realise that we have to walk a further two kilometres out of town to reach our destination, the Gîte Du Pontet that is located next to the town's campsite. The last hour of the day's walk is really tough and we are pretty much on automatic pilot as we finally find the gîte and, with a little help from the friendly manager, book ourselves in and gratefully slip off our rucksacks and tuck in to a refreshing beer.

We may be aching and exhausted, but Mont Blanc has been witness to far greater tragedies than Rupert's blister, which puts things into some considerable perspective and makes one realise that there is a much darker side to this mountain. On 3 November 1950 an Air India flight, the *Malabar Princess*, crashed into the French side of Mont Blanc, tragically killing all on board. Bizarrely, one of the engines from the wreckage was found some 39 years later on the surface of the Glacier des Bossons and a second engine from the same plane 58 years later on the same glacier. Sixteen years after the *Malabar Princess* crash, in 1966, another Air India flight crashed in almost exactly the same spot on Mont Blanc, again killing all those on board. The only survivors were a number of monkeys that were being transported in the cargo hold for medical experiments and were found wandering around in the snow by rescuers.

In fact, in August 2012 just a couple of weeks before Rupert and I set out on this trip, a jute bag of diplomatic mail, stamped 'On Indian Government Service, Diplomatic Mail, Ministry of External Affairs' was discovered by a mountain rescue worker and handed in to local police in Chamonix. Among other things, the bag contained still-legible copies of *The Hindu* and *The Statesman*, dating back to January 1966. Even more recently, a French climber was approaching the summit of Mont Blanc when he found a metal box containing precious gems, rubies and emeralds worth over £200,000 packed neatly in sachets marked 'Made in India' and again these were clearly thrown up from one of the Air India plane crashes. The climber, an honest man, carried the box of treasure back down with him and handed it in to the local gendarmes.

Stories like this are not that uncommon as glaciers have a nasty tendency to swallow things up (more often than not people) and then many years later the appropriately named glacial 'tongue' will unceremoniously spit them out further down the line. One such story again concerns the Glacier des Bossons and an expedition led by Dr Le Chevalier Hamel, a Russian scientist and M. Selligue, a French optician, who set off on 18 August 1820 to climb Mont Blanc. This was 34 years after the aforementioned Michel-Gabriel Paccard and Jacques Balmat had first climbed Mont Blanc and since that time 14 other parties had reached the summit, surprisingly without any accidents. Hamel and Selligue led a party of 16 people, which included 12 local guides, but after hitting bad weather Selligue and two of the guides decided to head back down to safety while Hamel and the others pressed on for the summit.

It was a brave but foolhardy decision in the conditions and as they climbed higher the snow gave way and the whole party were swept down in a great avalanche. Three of the men, the guides Pierre Balmat (no relation to Jacques Balmat I believe), Pier Carrier and Auguste Tairraz, fell into a crevasse and were considered lost forever. However, some 40 years later our fickle friend the Bossons Glacier began to spit out some unusual objects including fragments of clothing, various items of expedition equipment and most of all, human remains. Apparently, a skull with hair attached, an arm, a hand and a foot all emerged from the tongue of the glacier. Word has it that Joseph-Marie Couttet, one of the surviving guides who was in his seventies at the time, could identify each object and notably said in the police report: 'I never thought that I once more should grasp my old friend Pierre Balmat's hand.' It seemed that the unfortunate guides had travelled ten kilometres over a period of 40 years from the crevasse where they had first vanished to the point where they eventually reappeared – disgorged by the Glacier des Bossons.

The dormitory-style gîte is great and thankfully also quiet with only a handful of other walkers in residence, including a couple of girls who are admirably cooking up some food on a little camping stove outside. Not being quite so hard-core, the manager phones over to the campsite restaurant to check that they are still serving food as it is getting pretty late in the evening. Without even time for a shower we have to hotfoot (or hopfoot in Rupert's case) our way through the campsite to the restaurant where a superb dinner of beef, pasta and salad and fruit salad is waiting for us, which we wash down with some red wine, all for around €12. There are a few other people eating and we notice a young guy walking between the groups checking out information on the next stage of the TMB. We are impressed with his effortless networking skills as he gathers information about the route, the weather, the accommodation options as well as no doubt sizing up potential walking companions. He gives us a wide berth, perhaps aware that we have just arrived and are barely able to communicate. After eating we head back and have a welcome shower before heading into our bunk beds for some much needed sleep.

Five

As well as the expected climbers, adventurers and explorers Mont Blanc has attracted a number of other less expected visitors hoping to make their mark on this great European high point. Literally in the case of the Chilean-born, Danish-resident 'guerrilla artist' Marco Evaristti who was arrested on 6 June 2007 for attempting to paint the summit red, apparently with a biodegradable dye made from crushed raspberries. Being half Danish myself I hold some admiration for this guy's chutzpah (that's my Jewish half coming out there) for the audacity of his deed and for what in the right weather conditions could have broken the record for the world's biggest raspberry smoothie. Not content with attempting to paint the mountain red, two days later Evaristti climbed Mont Blanc again and draped part of the summit in red fabric and planted a six-metre pole with a flag reading 'Pink State'. When asked to justify his actions, Evaristti said he was protesting against the day trippers and 'casual climbers' who polluted the mountain with their debris and litter. I guess that this litter doesn't include several hundred square metres of red plastic-covered fabric and a few litres of red dye then. Other notable 'works' by our friend Evaristti included an event in which he invited friends to a dinner of meatballs made from the fat of his own body extracted by liposuction, and an art installation that invited the public to put live goldfish through a food blender. One of them dutifully did, killing a few fish in the process. Following the predictable charges of cruelty, a court in Denmark ruled that the fish were not treated cruelly as they had not, according to the judge, faced 'prolonged suffering'. This, after evidence given by an expert witness from Moulinex. You couldn't make it up, as they say. Actually, what impresses me most about Evaristti (OK, the only thing that impresses me about him) is that he seemingly managed to climb Mont Blanc twice in the space of three days, his pockets jam-packed with fabric, paint, meatballs, flagpoles, fish, raspberries and who knows what else.

But within all this tomfoolery there may perhaps be some validity about Evaristti's Mont Blanc protest. Local politicians have apparently raised concerns about the increasing level of pollution on the mountain, claiming that Mont Blanc could well be renamed the 'grey and yellow' mountain as a result of the amount of rubbish that is often strewn the length of some of the most popular routes and the urine soaked glaciers turning an unpleasant shade of yellow. Suggestions of paid for permits or licences to climb the mountain have been made, with mixed reactions from the climbing community, but my sense is that at some point this is an environmental issue that needs to be tackled, popular or not.

Guerrilla artists we are not, but Rupert does have a sketch pad and watercolours in his rucksack and is itching to use them, although he won't be painting Mont Blanc red this morning as he is still concerned about the state of the ugly blister on his heel and how this will affect his progress. Still, he patches it up the best he can with a couple of plasters and we hope for the best. Today we have a tough ascent of 1,316 metres before reaching our planned destination of Les Chapieux, a distance of 18 kilometres. We both slept pretty well, unusually for me, and are up at 7.00 a.m. and eating a breakfast of toast, jam and coffee downstairs in the gîte's dining area. There are a few fellow walkers at the table and we chat to the guy we saw last night in the restaurant and sure enough he checks out our plans: where we think we might be staying, the possible route options, overall height gain and loss over the course of the day, the weather outlook and you can almost hear the bells and whistles of his hiking brain calculating and recalculating his choices as he assimilates new data in his route plan.

We set off from the gîte and the initial route follows the river and leads gently upwards through a lightly wooded area. It is a fresh morning and the sun is just beginning to break through the light clouds, bringing some welcome warmth. Unlike yesterday, there are many more walkers on the trail having more or less set off together, some from the campsite and the gîte and others from Les Contamines, the small town we briefly passed through last night. The gradient of the path quickly increases as we rise steeply up the side of a wooded ravine, the valley floor falling away to our right. The path is made up of large, Roman-laid slabs, many worn to a stony shine, and I guess this once served as a mule track as goods were transported from valley to valley between settlements. This is soon confirmed as we cross an ancient Roman humpbacked bridge, the Pont de la Téna, and we can look down and see where the tumbling torrent of water has carved deep clefts into the bedrock over the years. We pick up height all the time and the walking begins to become much more strenuous. As we look back we can see the track that snakes up from the valley floor, interspersed with walkers who look like a little trail of worker ants in the distance, carrying their precious loads on their backs and we are surprised by how much height we have gained in just an hour or so.

It is strange to think of this constant trail of people who, in the summer months, both walk around and climb over Mont Blanc in their thousands. Back in the day, in 1760, the price that Horace-Bénédict de Saussure put on the summit galvanized a cast of characters in pursuit of the purse and to be the first to conquer the mountain that had become an increasing object of Saussure's desire, an obsession no less. 'It had,' he wrote, 'become with me a species of disease; my eyes never rested upon Mont Blanc... without my undergoing a fresh attack of melancholy.'[3] A key player in this cast of alpinist characters was Marc-Théodore Bourrit, a contemporary of Saussure and Precentor of Cathédrale Saint-Pierre Genève. In his fine book *Killing Dragons: The Conquest of the Alps*, Fergus Fleming describes Bourrit as 'an extraordinary man... artist, singer, womaniser, snob and interminable raconteur... he was an endearing coward with a genius for self-promotion'.[4]

Bourrit referred to himself as the 'Historian of the Alps' and the 'Indefatigable Bourrit' and the only seemingly redeeming characteristic within all this rather flamboyant grandiosity was his somewhat obsessive fascination with the Alps, an obsession that he shared with Saussure who, in all other regards, was his polar opposite. There is no doubt that although pompous and vain, Bourrit was passionate about the mountains, a passion that he sought to share with others through his paintings, writings and exhaustive lectures, but despite his extensive and self-professed knowledge of the Alps the one thing that he couldn't do – to his great chagrin – was actually climb them. In many ways Bourrit seems something of a comedic character. Despite numerous attempts at climbing in the early 1800s he was always, as Fleming states, 'thwarted by three deliberations; he dreaded cold, he disliked rain and he suffered from vertigo'. You might think these were reasons enough to deter any would-be mountaineer but perhaps it says something for the 'Indefatigable Bourrit' that he would not allow a simple thing like a fear of heights to stand in his way. He was often seen striding about the lower valleys wearing a red cloak and carrying a large red umbrella.

Fleming recounts an amusing story about one of the few occasions that Bourrit did actually climb a 'hill' and having set up his easel and watercolours noticed after a while that he had sunk waist deep into the snow and had to be dug out. In 1775 Bourrit managed to climb Le Buet, one of the smaller 'peaks' that sits at a mere 3,096 metres, and his apparent delight being such that be broke into a rousing round of song and dance. As a result of his excited performance, a great flat rock on Le Buet was christened 'La Table au Chantre' in his honour and still preserves his memory to this very day. Poor Bourrit, he may have been the object of some disdain and mockery by his fellow Alpine explorers, but I quite like the sound of this colourful fellow and as we stride along the TMB, me with my staff and floral trilby and Rupert with his sketch pad and watercolours, I like to think that we perhaps evoke just a little bit of the spirit of Alpine eccentricity.

As a consequence of Saussure's proffered reward and barmy Bourrit's fervent endeavours, a number of failed attempts had been made upon Mont Blanc's summit but it wasn't until 1783 that Bourrit declared that he was going to climb the mountain himself, in what could at best be described as a bout of foolishness given his rather limited mountaineering credentials. With the aim of securing some sense of scientific credibility to his venture he managed to persuade (after being rejected by several others) Michel-Gabriel Paccard to accompany him. Paccard was a young Chamonix doctor and something of an amateur scientist with a particular interest in botany and minerals and for some time had harboured his own desires to climb Mont Blanc, even scouting possible routes to the top during his plant collecting expeditions.

On 15 September 1783 Bourrit, Paccard and a number of local guides set out on their ascent, which perhaps unsurprisingly ended in failure. Bourrit, true to form, claimed it as a grand failure in which they bravely battled the elements that ultimately defeated them. Paccard, in his own account, wrote that 'we arrived at the glacier… Mont Blanc was covered with clouds and M. Bourrit did not dare go on the ice'.[5]

Suffice to say, after this attempt Bourrit and Paccard did not renew their mountaineering partnership. Two years later, in 1785, Saussure made his own attempt on the mountain, in the company of the 'Indefatigable Bourrit' whom he felt forced to reluctantly take with him as a mark of respect, and again this ended in failure and left some bad feeling between the two men. Clearly, the choice of one's companion is an important consideration, be it for a climb, walk or indeed any venture that requires two people to spend a significant amount of time in each other's company. I was very pleased that Rupert was able to join me on this walk and his enthusiasm from the very point that we shook hands on this challenge has helped make it happen. We have known each other for many years and there is an ease in our relationship that comes with familiarity and a comfort in each other's presence.

This eighteenth-century Alpine space race finally reached its conclusion on 8 August 1786 when Paccard, in the company of a local guide named Jacques Balmat, fought his way through high winds, altitude sickness and snow-blindness to eventually reach the summit of Mont Blanc after climbing for over 36 hours. Balmat was undoubtedly the key to this success. Descriptions of Balmat give a sense of a determined, feisty and physically strong character who, like so many others of the time, had begun to obsess about the idea of one day conquering the mountain that loomed gloatingly over his Chamonix home. During the months preceding his success with Paccard, Balmat undertook a thorough reconnaissance of Mont Blanc, climbing Le Brévent many times to get a good view of the various slopes and possible routes to the summit.

Most significantly, on a solo summit attempt in June 1786, Balmat was the first to spend a night in the open at high altitude, a feat that no one else had thought possible. In his account of this experience Balmat writes: 'My breath was frozen and my clothes were soaked... Everything was dead in this ice-bound world and the sound of my voice almost terrified me. I became silent and afraid.' Amazingly, Balmat then spent a second day and night on the mountain before deciding to give up and go back down.

Fleming in *Killing Dragons* recounts the story that on his way back down Balmat met a party of three guides undertaking their own assault on the summit (on the pretext of looking for goats). Balmat quickly rumbled their true intentions and could not bear the idea of them reaching the top and claiming Saussure's reward without him. The three men duly asked Balmat to join them to which he agreed. But being a dutiful husband and having promised his wife that he would be away for no longer than two days, he ran down the mountain and back to Chamonix, told his wife that he was back, grabbed a bit of food and promptly took off again. He caught up with the others two hours later for what turned out to be another unsuccessful attempt and when Balmat eventually returned home, exhausted and sunburned, he collapsed and slept for 24 hours after being up on the mountain for close to five days.

So yes, Balmat was made of strong stuff and was determined to be the first to conquer Mont Blanc. But like Bourrit, Balmat felt he needed a man of science to give credibility to his conquest and with this in mind he approached Paccard, who needed no persuasion to make the attempt. So on 7 August at 5.00 p.m. the two of them set off, without guides, on what was to be an epic moment in the history of alpinism. The only real first-person account of this climb was recorded many years later by the novelist Alexandre Dumas, who met with Balmat and persuaded him to tell his story of their climb. Of the moment when Balmat reached the summit, ahead of Paccard who was struggling to continue, he said:

> *I felt as if my lungs had gone and my chest was quite empty... I kept walking upward with my*

> *head bent down, but finding I was on a peak which was new to me I lifted my head and saw that at last I had reached the summit... I was the monarch of Mont Blanc... I was the statue on this unique pedestal.*[5]

Paccard and Balmat's conquest of Mont Blanc was a historic achievement in the history of mountaineering. As C. Douglas Milner wrote in his book *Mont Blanc and the Aiguilles*: 'The ascent itself was magnificent; an amazing feat of endurance and sustained courage.'[6] Indeed, the bravery, perseverance and sheer endurance of these early pioneering mountaineers like Paccard and Balmat and the many others who sought to capture the Alps is quite astounding. Without the knowledge, equipment and support that are now taken for granted, they forged a path (literally and metaphorically) for others to follow.

Soon after their return, Balmat travelled to Geneva to claim Saussure's prize and an account of the ascent was written up by our old friend and 'Historian of the Alps', one M. Bourrit, who was intensely resentful and jealous of Paccard and Balmat's success. Bourrit was mischievous in his written account of the ascent, downplaying Paccard's achievement and implying that he had to be dragged to the summit by Balmat. Not helped by Bourrit's provocation, there was much rankling between the key players in this drama, claim and counter-claim, and it is a great pity that a degree of the celebrated success of this conquest was lost in the enmity of the post-match analysis, so to speak. Paccard made no claim on Saussure's prize, stating that he did not need the money, but this may have furthered the impression that somehow he was second fiddle to Balmat's grand finale.

In a final endnote to this story, Paccard later married Jacques Balmat's sister and became a Justice of the Peace. Balmat himself died in 1834 after falling from a cliff in the Sixt valley while prospecting for gold. The 'Indefatigable Bourrit' was finally pensioned off by Louis XVI (not bad since it seems he was absent from cathedral duties for much of the time while gallivanting in the mountains) although this seems to have ultimately paid off when he was named 'Historiographe des Alpes' by Emperor Joseph II.

As for Saussure himself, perhaps the chief protagonist in this whole drama, he finally reached the summit of Mont Blanc in the morning of 3 August 1787 in the company of Balmat who had offered his services as a guide. After all his years of obsessing over Mont Blanc and dreaming of the time when he would one day conquer this great object of his perennial desire, the actual moment seemed something of an anticlimax. In his journal Saussure wrote:

> *... the arrival was no* coup de théâtre *– it did not even give me all the pleasure one might have imagined. The length of the struggle, the recollection and the still vivid impression of the exertion it had cost me, caused me a kind of irritation. At the moment that I trod the highest point of the snow that crowned the summit I trampled it with a feeling of anger rather than pleasure.*[18]

I guess anticipation is everything, as they say.

Back on the trail of the TMB, Rupert and I are anticipating the moment that we get to the end of our own ascent for the day and seek to amuse ourselves by trying to guess the occupation of the guy we were chatting to at breakfast. Teacher? Accountant? Detective? Doctor? After a while we stop for a breather and a snack of bread and apricots to keep our energy levels up and catch up with our mystery friend who actually turns out to be a surveyor from Cardiff called Richard. We chat together for a few minutes and also with the two girls that we saw cooking up their dinner last night outside the gîte and as we do so I slip off my rucksack and stand it up on the ground by the side of the path with my stick propped against it and my hat hooked over the top of the stick. Moments later I hear a shout from Rupert and turn to see, in agonising slow motion, my rucksack toppling over and my precious floral trilby roll over the edge of the path and expertly spin on its rim down the steep side of the valley towards the river at the bottom.

Rupert seems pleasingly entertained by the episode and makes a few sympathetic comments about how no doubt I can pick up another hat along the way somewhere, but he has underestimated the strength of my attachment to my little floral friend and after a quick risk assessment of the situation I disappear over the edge myself, in a slightly more considered fashion. There is going to be no mourning for my hat just yet. The slope is so steep I have to slide down on my backside and while rather undignified I reassure myself with the thought that Saussure, Paccard, Balmat and perhaps even Bourrit would have undoubtedly gone after their own hat in similar circumstances (or more likely sent one of their guides down for it). Anyway, I manage to bum-slide my way down to the bottom of the slope and then clamber over a few rocks to find my hat miraculously wedged behind a stone about twelve centimetres from the stream. Victorious, I grab hold of it and then make my laborious way back up the slope by moving in a strange sideways crab-like manner, occasionally slipping and having to grab onto clumps of grass to stop myself sliding back down to the valley floor. Looking back, this was one of the more exhausting ascents I had to make during the course of this trek and one that I had not planned for. Still, safely back on the path, I dust myself down, pull my hat firmly onto my head and we continue on our merry way.

We push on ever upwards and pass by the tempting Refuge de Col la Balme that sits at the foot of the Aiguille de la Pennaz (the Pennaz needles), a dramatic collection of jagged peaks of around 2,688 metres that dominate the landscape with their craggy grey spikes. We consider stopping for a break in the refuge, but it looks closed and so press on with a long hard slog that seems to go on forever. The path here is rough and stony and zigzags its way ever upwards, its twisty hairpins compensating for the steepness of the ascent and soon we leave the treeline and enter the basin of the Plan Jovet, which is fed by streams that cascade their way down from a hanging valley somewhere above. From here the path leads us onto the Plan des Dames, marked by a large cairn – a hand-built pile of rocks – that word has it is the spot where an English woman was killed in a storm. I suspect that we might see a few of these on our way around the TMB. Stone cairns like this are constructed for many reasons, mostly as route markers where the direction of a path or junction may not be clear, but as in this case they are sometimes built as a memorial. Mountain summits or passes are also invariably marked by a cairn, walkers often adding a stone or two as they pass by. Whatever their particular symbolic significance, these stone constructions add a human touch to the trail, an indication of the timeless passage of people along these well-worn mountain ways.

High above us, we catch our first, relieved glimpse of the col and before long we hit the snow line and find ourselves walking at points through little dotted islands of deep crusty snow, along with a strong, cold wind that suddenly picks up and periodically blows in so fiercely that it nearly knocks me off my feet. Eventually, we get to the saddle of Col du Bonhomme, which sits at close to 2,400 metres, where a small cluster of hikers have gathered to rest, take in the surroundings and shield themselves from the cold in a small wooden shelter. This is the highest I have ever been in my life and the fierce wind and sudden drop in temperature is something of a reality check as I realise that this is a serious business. Other walkers around me are pulling out various items of protective clothing and as I rummage futilely for my woollen hat and gloves I wonder if I am adequately equipped for this trip. I find my fleece and quickly pull it on, but feel myself getting a little anxious about safety with the realisation that conditions up here can change so quickly and dramatically (dare I say it, at the drop of a hat), and my anxiety is not helped by noticing that a lucky charm attached to my rucksack (made and given to me by Jess before my trip to the Pyrenees a few years ago) has broken off. Damn! I am not superstitious, but all the same the timing doesn't feel good.

Eager to keep warm we press on and pick our way through a wild landscape of scree and rock interspersed with white patches of snow and the occasional precarious water crossing.

At one point a friendly marmot pokes its head out of a hole just by the side of the path and indulges the passing walkers by posing for photographs. It's a friendly but clearly narcissistic marmot that has been seduced by the culture of celebrity and it revels in the attention proffered by attendant walkers who stop in groups and take photographs like some kind of Alpine paparazzi. The marmot, half beaver and half giant guinea pig, poses obligingly and Rupert gets some fine photos that no doubt he can syndicate to the Sunday papers when we get home. Being usually very shy creatures, I wonder what the other marmots think about their fancy friend, hobnobbing with the humans with reckless abandon. We also briefly see an ibex in the distance, its large curled horns distinct against the sky as it stands up high on an outcrop of rock looking down at us with a cautious interest.

At about 3.30 p.m. and after about six and a half hours of walking we reach the Refuge de la Croix du Bonhomme where we have decided to rest up for a bit of tea and cake for which it seems to be famous according to word on the TMB grapevine. It's a delightful place, sitting at 2,443 metres with stunning views all around and Rupert and myself are somewhat entranced (or possibly exhausted) as we sit and tuck into delicious cake and hot chocolate. From here there are a couple of options: to continue on the main TMB route down to Les Chapieux or take a TMB variant over the top via Col des Fours to the Refuge des Mottets. We are warned that the variant option should not be taken if there is any snow on the ground or if the weather seems unpredictable and there is much discussion going on within the refuge about whether to 'go down' or 'go over' and our friend Richard expertly works the room to establish what everyone else is doing before he makes his move. Meanwhile, Rupert and I have chosen a radical third option, which is to hole up and stay here in Bonhomme for the night. It's cosy, warm, relaxed and very chilled out. We've walked for seven hours and we are in a terrific spot high up in the Alps. Frankly, why go anywhere? Richard is temporarily thrown by our left field move and while initially saying that he will stay here with us, suddenly attaches himself effortlessly to another group and heads off down to Les Chapieux.

And so it is that we spend a remarkably pleasurable late afternoon idling around in the charming Refuge de la Croix du Bonhomme, drinking coffee, writing our journals and chatting to other hikers. The place is somehow quite timeless and very peaceful until later when waves of walkers and even a few mountain bikers blow in with little blasts of activity – all carrying stories, information and advice for the next stage of our respective journeys. As evening descends the refuge keeper and his small team put candles out on the tables and in lieu of any electricity we sit among the warm, atmospheric glow of candlelight as dusk sets in, bringing with it a spectacular sunset that sends us all scampering outside with our cameras to capture the moment. Later we have a fabulous dinner of lamb and pasta, followed by cheese and cake and chat to a charismatic and entertaining guy from Guatemala and his German partner. It feels strangely magical to be sitting enclosed in this isolated, solar-powered building, talking to fellow walkers in the soft candlelight; someone playing a guitar, some playing chess or Scrabble, others planning their route tomorrow with large maps spread over tables, all chatting in a lovely linguistic array of French, English, Spanish, German and Italian.

I feel softly cocooned in this little island in the sky, sheltered from the gathering darkness. There is something very evocative about the experience of shelter – it's a felt experience, a physical sensation of safety and security, a counterpoint to the tinge of anxiety I felt earlier, high up on the crossing. It makes me think of Balmat's first night on the mountain, perched upon the hard snow with no cover, his frightened voice dead against the freezing mountain air. It makes me think of childhood dens, camps in the woods and tree houses where I would hole up for hours at a time. It makes me think of my work as a play therapist, the playroom a place of both physical and psychological safety for the child. It makes me think of nights in our camper van, wrapped up in the warm while a storm rages outside, the rain like bullets on the hard roof. Unable to resist, we order another carafe of red wine and allow ourselves to be slowly and deliciously enveloped by the delightful Refuge de la Croix du Bonhomme.

Six

The first woman to climb Mont Blanc was Marie Paradis, a young maidservant from Chamonix, who in 1809 was more or less dragged to the summit by Jacques Balmat and a couple of other guides who had bumped into Paradis as they made yet another ascent of the mountain. It would seem that she had not especially intended to climb Mont Blanc that day (Why would you? It seems she only went out for a little walk.), but had been beguiled by Balmat's wolfish charm and the prospect of good earnings if successful. After camping overnight on the Grand Mulets the small group struck out for the summit and unsurprisingly the unprepared Paradis struggled badly with exhaustion and snow-blindness and pleaded with Balmat to go more slowly or else leave her behind. Never one to compromise and not a little ungallantly, Balmat and his associates took poor Marie by the shoulders and hauled her across the Grand Plateau and up to the summit. She was in a bad way, but somehow they got her back down alive to Chamonix where she was something of a cause célèbre and became known locally as 'Maria de Mont Blanc'. As promised by Balmat (and deservedly considering the gruelling experience she had been through), Paradis did well financially from the climb, setting up a small tea shop and providing refreshments for visitors and passing Alpine explorers. The written account Paradis later gave of her ascent of Mont Blanc was in some contrast to the generally immodest and lavish accounts proffered by her climbing predecessors of the period. She simply writes: 'I climbed, I could not breathe, I nearly died, they dragged me, carried me, I saw black and white, and then I came back down again.'[7]

One can't help but admire this unassuming woman, who in the midst of all this pre-Victorian fervour for Alpine adventure was quite literally propelled to the heights of Mont Blanc in the most unexpected of circumstances. It was not until some 30 years later that another woman, the aristocratic Henriette d'Angeville, led by the experienced guide Joseph-Marie Couttet, conquered the mountain and was fittingly welcomed on her victorious return by Paradis herself, then in her sixties. D'Angeville has sometimes been referred to as the first woman to reach Mont Blanc summit 'under her own strength' (Paradis was more or less carried to the top by Balmat and his guides), but this seems harsh in my view. Bravo Mme. Paradis, that's what I say. On the way up the mountain d'Angeville was reported as saying to Couttet: 'If I die before reaching the summit take my body and leave it there; my family will pay you for fulfilling my wishes.' 'Be easy in your mind,' he replied, 'living or dead, to the top you shall go.'[8] Upon reaching the summit Couttet said to d'Angeville, 'Now you shall go higher than Mont Blanc' and the guides lifted her up as high as they could.

Meanwhile, back at the Refuge de la Croix du Bonhomme, Rupert and I have slept very well and neither of us can recall a single, sleep-disturbing snore in a dormitory of close to 50 grizzled trekkers, which is a feat in itself worthy of some celebration. The dorm has been built underneath the refuge like some kind of giant earth burrow and we emerge blinking into the bright early morning sunlight, rubbing our sleepy eyes like little marmots waking from a winter's hibernation. The refuge itself is a gentle bustle of activity as people eat breakfast, study maps, compare routes, pack their rucksacks, queue for the minimalist shower and generally make preparations for the day's walking ahead.

Mindful of his own day's walking ahead, Rupert last night took the opportunity to talk blisters with several of our fellow walkers and has discovered the joys of Compeed Blister Plasters, seemingly magical 'second-skin' patches, and has even managed to get his hands on one, which he excitedly applies to the ailing heel in question. So forget euros, sterling or the Swiss franc – it seems that the Compeed is the underground currency of the long distance hiker. Little groups of hikers cluster furtively in the dark recesses of the refuge, craftily palming the little round plasters while nervously glancing the other way, as if the TMB police might bust them at any moment. We make a note to stock up on the little fellows at the next opportunity just in case we need to bribe our way into Switzerland later in the week.

It is a gorgeous morning, the clear blue sky heralding another fine day ahead which means we can take the TMB variant and go 'over the top' across the Col des Fours and then drop down into the valley beyond. Our planned route for the day actually takes us across the border into Italy, which is quite exciting, and then on to our chosen rest point for the night, the Rifugio Elisabetta. This is a walk of around 16 kilometres with a total of 1,010 metres ascent and 1,270 metres descent, a long enough stretch for our third day, so after fortifying ourselves with some toast, jam and several cups of strong coffee we prepare to head off – invigorated but with a tinge of reluctance to leave the comfort of the refuge. As we are about to set off we spot another ibex, or perhaps the same one, standing regally on a ridge about 70 metres away. It is an impressive looking beast and there is a proud aloofness to its presence as it maintains a safe enough distance from the human commotion around the refuge while still communicating a strong sense that it is we who are the intruders here in the uplands.

Under the watchful eye of our ibex friend we leave the refuge and make the short way back up the slope to the saddle of the Col des Fours, which sits at 2,665 metres, so I have already broken my lifetime height record that was set only yesterday on the Col du Bonhomme. In fact, this is the highest point along the TMB, so the new record should stand for a while. Patches of gleaming snow, large and small, dot the grey rock as if some kind of randomly designed patchwork quilt has been casually strewn across the landscape and although there is a chill in the morning air the whiteness of the snow dazzles in the bright sun. We pause for a while to take in the scene while some of our fellow walkers continue up to the 2,756-metre summit of the Tête Nord des Fours for the 360-degree view afforded by this optional, extra bit of exertion. Rupert and I are happy where we are; the views are stunning in whichever direction we look and it is hard to think they will be bettered, so we conserve our energy and savour the moment.

Mont Blanc itself, a shining dome of snow and ice is visible to the northeast and this is our best view so far of the object of our hiking ambition. To think that we are going to walk all the way around this mountain, which hangs in the air like a snow-capped island in a sea of blue and grey, seems quite absurd. The view is simply breathtaking. Like theatre staging, the dome of Mont Blanc protrudes from behind the ridges and needles of the surrounding massif; curtains of rock that allow just a tantalising glimpse of the main act. The stark contrast of the white and grey is softened in places by a foreground of green that carpets great swathes of ground, and in the early morning sun, the sharp relief of the landscape is picked out by the chiaroscuro contrast of light and shade, giving a rich texture and depth to the overall scene.

When she was young, I used to make up bedtime stories for my daughter about the sky giants, great lumbering figures that inhabited the in-between place where the sky ended but before space began. With great rusting machines the sky giants would grind out clouds and with a sharp puff of their icy breath send the clouds skidding and thudding across the sky. The cracks and rumbles of thunder were not Thor-like Asgardian expressions of anger, but instead were the giants at play as they tossed around the sun and the moon for fun and scattered the stars like marbles across the horizon floor. Occasionally, when the fancy took them, the sky giants would bend down and push enormous wizened fingers through the clouds to the ground and carve patterns in the earth, like young children playing in the sand. When I look out now at the scene before me, I can't but help think of the sky giants. It is as if the same giant that spread the snow-encrusted quilt across the landscape had also taken up some tweezers and idly picked at the seams and folds of the earth, tugging at the very fabric of the rock and pulling it up into pointed peaks and sharp ridges to produce this magnificent range.

These moments, these places create a sense of insignificance and somehow put one's place in this world into some kind of context. In the face of such wonder it is hard to hold onto the niggling array of worries and anxieties that constantly buzz around my head like a swarm of neurotic bees. Has the camper van's MOT run out? Can I mark those dissertations in time? Should we paint the front door? Will I survive 50? What's that pain in my chest? Am I happy at work? Am I happy? Somehow all these troublesome intrusive thoughts simply fall away, drift away with the cool mountain breeze. Perhaps they float up to the in-between world of the sky giants who pluck them from the air and pack them into their great, grinding cloud machines that turn them into beautiful crimson skies of cirrus and stratocumulus. Aldous Huxley once said that his father considered a walk among the mountains as the equivalent of churchgoing. I am not a religious man – far from it in fact – but there is something deeply spiritual about these places. Not in any kind of external way but more in the sense of a connection with some kind of inner tranquillity, a place of calm contemplation. In a strange kind of way, nothing seems to matter when you find yourself transported into this kind of magical wonderland.

As well as psychologically, I also get a physical sense of my place in the world from this lofty perspective. The contours of the Western Alps ripple and buckle all around me. To the north, if my eyes allowed it, my gaze would take me over towards Lake Geneva in Switzerland, then through France and Germany and across the North Sea to Norway. To the east: Italy, Slovenia, Romania, the Black Sea and the southern foot of Russia. Southwards: the Mediterranean and the warmer climes of Tunisia and North Africa, and looking west: France, the Atlantic and eventually Nova Scotia at the south-eastern tip of Canada. This is my place on Earth: tiny, irrelevant, inconsequential… just a bit-part actor in the great global drama that is played out upon this vast stage of ours. Time itself takes on something of a different meaning in this kind of space, the relief of this great range providing a frosted window into the past, the physical formation of the mountains a story of their own creation.

Essentially, the Alps are part of a tertiary collision belt and it is almost possible to imagine, all those many millions of years ago, the slow-motion collision of the African and Eurasian tectonic plates that pushed, thrusted, folded and lifted the protesting rock ever upwards as one plate slipped under the other. Old sea floors from earlier eras were lifted high into the air, buckled and twisted, leaving the fossil record of tiny sea creatures deposited upon the newborn peaks, a puzzle for early explorers who saw this as evidence of the Great Flood, the Alpine peaks rising like Ararat from the receding water. In fact, prior to the twentieth century the idea that mountain ranges were the product of the continental-scale movement of vast tectonic plates was unheard of and any such idea would have been considered potentially blasphemous within the dominant religious ideology of the time.

A few people dared to be curious, one of these being the Anglican churchman and philosopher Thomas Burnet who, in 1681, wrote a challenging text of new ideas called *The Sacred Theory of the Earth*. Burnet, who was quite taken with the Alps, introduced a touch of rationality to the debate and questioned both the fact that the mountains were not included within the biblical account of creation and that the Great Flood could not possibly have covered mountains of this height – even if it had rained solidly for 40 days. Burnet's hypothesis was that the Earth was completely smooth at the point of creation with water trapped within its core upon which the land floated. As time passed, according to Burnet, the crust of the Earth was cracked and fragmented by the relentless heat of the sun and the water eventually broke free to cause the deluge. This epic catastrophe led the rock and crust of the Earth's surface to be broken and fragmented and when the water eventually receded the aftermath was, in Burnet's words 'a world lying in its rubbish... wild, vast and indigested heaps of stone... the ruins of a broken world'.

So these then were the Alps; piles of debris – remnants of the cracked earth – left scattered by the great waters. A little like the tectonic plates themselves, these ideas form the critical points of collision, friction and overlap between the worlds of science and theology, reason and romanticism, beginning with the gentle questioning from people like Burnet, who attempted to accommodate and integrate his ideas into his existing belief system, and much later of course with people like Charles Darwin, who in the nineteenth century created something of a tipping point and sent seismic shivers around the Christian world with his ideas of natural selection. Our friend Saussure thought that the Alps were the result of some great subterranean explosion – not too far from the truth as it happens – and later theorists and geologists like James Hutton and Charles Lyell began to lay some much firmer theoretical foundations for succeeding geologists in the nineteenth century to pick up and run with. In fact it was Alfred Wegener, a German geophysicist and meteorologist, who advanced in 1912 the theory of continental drift and the notion that these shifting land masses were somehow accountable for the phenomena of mountain formation. Wegener's ideas were initially given short shrift by the scientific community, but over the following decades were gradually accepted and then finally confirmed in the 1960s. Poor Wegener himself froze to death in 1930 in a blizzard in Greenland, one storm that the meteorologist was clearly not able to predict, and so sadly never saw the fruition of his early theories.

The Alps form just one part of a great tertiary orogenic belt of mountain chains, the Alpide Belt, which stretches through southern Europe and Asia and from the Atlantic all the way to the Himalayas. The ripple effect from this massive, tectonic collision created other lesser ranges closer to home, like England's South Downs for example, distant echoes of grander forces at work. Of course, the Alps have been shaped by many ice ages over subsequent millennia and their current formation is relatively recent in the grand scheme of things. And here I am, contemplating my fiftieth birthday and worrying that no one will turn up to my party because really I haven't got any friends and everyone will have something better to do on that day anyway. I wait a moment for that particular thought to be gently lifted away and turned into a little fluffy grey cloud by my sizable friends upstairs.

We reluctantly leave the high point and head south-eastwards to begin the long descent down the valley to the basin of the Plan des Fours, some several hundred metres below. The path is made up of loose, grey shale and patches of slushy snow and ice, although we begin to lose the snow along with the height as we zigzag our way ever downwards. We have to carefully pick our way along the track, the loose shale and scattered larger stones making it tricky at times and at least twice I slip and land on my back with a thump, my rucksack thankfully breaking my fall like a protective shell. Normally I am pretty sure-footed when walking and put my falls down to being somewhat distracted with thoughts of sky giants and what shape the cloud of an impending 50-year-old might look like. Rupert is both mildly amused and concerned by my carelessness, but reassures me that 'the thing is the thing' – presumably meaning that if I slip and break my leg I will just have to take it in my stride, or not as the case may be. Rupert himself is in Compeed heaven and verily skips along the trail and I wouldn't be surprised if he breaks out into a whistle at any moment, so grateful is he to be in a blister-free state.

Both ahead and behind us are fellow walkers making their way down to the valley floor and once again we get a sense of the number of people on the TMB; many more than we expected for this time of year. The trail continues to descend, steeply at times, and seems to go on forever. The green and grey textured, rippled slopes of the valley rise up on either side of us as we continue downwards and ahead is an impressive view of the Montagne de la Seigne that rises up like a great wall. The track takes us at first alongside and then eventually over a stream and past the deserted looking 'hamlet' of Les Tufs (1,993 metres) – in truth just a few deserted outbuildings. The path then joins a farm road and leads us by another small cluster of farm buildings, rather optimistically called La Ville des Glaciers, and after crossing the river again we gently climb following the course of the river until we reach the delightful Refuge des Mottets, where we stop for a breather and spot of lunch.

The refuge, a converted dairy, has a laid-back and rather atmospheric charm about it and nestles picturesquely at 1,864 metres at the foot of the Glaciers des L'aiguille, which rise dramatically overhead. The outbuildings have been converted into solid, functional dormitories – a veritable feast of original wooden beams and stone floors – and the main building/restaurant is a wonderfully welcoming place, full of traditional farming paraphernalia and odd agricultural artefacts that hang in abundance from the ceiling and walls, lending the space something of a museum-like quality. Only recently converted in 2010, the refuge has really held onto its dairy farm heritage and the whole place has a solid earthiness about it that is really quite captivating and very welcome after several hours on the trail. We sit outside drinking hot chocolate and eat the picnic lunch that we bought from Bonhomme. The place is quite abuzz with walkers milling around: resting, eating, planning and some even sleeping. These refuges may be basic, but they are functional, comfortable and well organised and provide good food, drink – pretty much all that a walker can desire – except perhaps a single room with en suite bathroom, Jacuzzi, newspaper, a glass of fine red wine and a piano. I don't ask for much.

Rupert is tempted to stop for the day here to soak in the atmosphere and do some painting. I am not sure, preferring to continue to the Rifugio Elisabetta for the night, which means we get to Courmayeur tomorrow, where perhaps we might take a rest day. Rupert in fact is much better at our adopted 'the thing is the thing' philosophy. He is more adept at staying in the moment than I am – happy to pause… stop… relish the experience of the here and now while I have an incessant and nagging urge to 'push on' and keep moving. It is as if I am caught in a constant state of restlessness from which I can only escape by being constantly on the move. Actually it is something of a family trait. During any gathering of my rather vast family you could probably power a small town through the kinetic energy generated by our collectively incessant pacing. Rupert is content to be still and I wish I had a little more of that quality.

Talking of my family, now and again as I walk my awareness temporarily alights upon the small, plastic vial containing my father's ashes that lies tucked away in the top pocket of my rucksack, alongside my orange survival bag, my camera and a crumpled bag of dried apricots. It feels odd, carrying him like this on my back, if indeed the grey-brown dusty remains actually constitute a 'him' as it were. For many years, from my early teens to my mid-twenties, it felt as if I carried my father on my back, burdened by his troubled, oppressive presence that weighed me down, not unlike my rucksack now. I walk forwards, but inevitably my thoughts take me backwards, stirred by the meditative repetition of my steps upon the glacial scree beneath my feet and the continual presence of the cathedral mountains above.

Seven

Allowing these thoughts to take me back, I recall that as a young child it was often exciting growing up with a father with such extreme mood swings, although I was too young to make much sense of it then. When he was manic, I would be swept away by the infectious spirit of his magical mood as he sang, danced, laughed and joked his way around the house like some kind of whirling Jewish dervish. When it was Christmas he would run around the house, whooping with joy like a young child. As a young child myself, I would sit on his knee, entranced, while he told me his favourite story about a boy falling down the toilet and being swept out to sea and swallowed by a great whale. And when he went on his regular walks 'around the block', with his wooden stick and eccentric hat, I would buzz around him just happy to be in his excitable orbit. As is so often the case with manic depression, my father could be an incredibly charismatic man: witty, erudite, charming, sometimes magnificent. To have been able as a young child to bathe in his radiant light, fizzing with luminescence was really quite something.

But a little like a lighthouse, every once in a while the light would move away or simply switch off, often for months at a time. A thick, treacly gloom would descend upon the house and my father would shut himself away in his study, the tap tap tap of the keys on his ancient typewriter and the thick, pungent smell of cigars the only sign of life from behind the unapproachable door. It is said that no man is an island, but during these times of depression my father certainly was remote and inaccessible, distant and unfathomable. He could be angry, unpredictable and hurtful and we had to weather the storms again and again, batten down the hatches and wait for the skies to clear.

I grew up in a rambling house in East Sussex, my father absent for much of the time due to his work as a consultant locum surgeon that took him all around the world, while my Danish mother struggled to make ends meet and keep the proverbial wheels on the family wagon. My father made good money as a highly skilled orthopaedic surgeon, but somehow did not quite grasp the concept of supporting one's family and so my mother brought us up more or less alone on her nursing salary. Being a large family this was no mean feat. All in all, my childhood could best be described as bohemian, perhaps bordering at times on the feral – in a good way. Imagine a cross between *The Famous Five* and *The Wasp Factory* (minus the burning sheep). That would be about it.

But I wouldn't change it for anything. Along with my elder brothers, I roamed freely around the local countryside; swimming in rivers, falling out of trees, sticking my head in wasp nests, blowing things up and merrily breaking bones and knocking myself out along the way. In fact, between us there were enough accidents and injuries to sustain several seasons' worth of *Casualty*: cue scene of older brother throwing two-metre bamboo spear into younger brother's eye. You get the picture.

This was back in the day before health and safety, before the rise of the 'worried well' and the 'helicopter' parents who hover over their children for fear that they might stub their toe on the new decking or get abducted by the child killers that lurk on every street corner. It's a sad thing that children don't play outside so much these days. Some research has suggested that close to half of all children today never play outside and according to the Children's Society, 43 per cent of adults believe that children under the age of 14 years should not be allowed to play outside unsupervised. When one weighs this up against the value of outdoor free-play the long-term consequences are pretty grim. Personally, I think it is important to experience risk – to a degree – to know what hurts and what doesn't hurt, to discover where the boundaries lie.

I remember one winter, when Jess was around seven years old, we had a particularly heavy and very rare snowfall such as we hadn't seen in many years – certainly not in Jess' lifetime. It occurred during a school day (yes, amazingly the school stayed open) and later Jess told me that they had not been allowed out to play in the snow at break-time. Instead, the teacher led them out in little clusters of twos and threes where they could bend down to touch the snow for a minute or so to see what it felt like before being led back into the 'safety' of the classroom. Well, you can guess I was furious; furious at this mind-numbingly corporate, neurotic and cowardly intrusion into my daughter's childhood; furious that the school in all its fevered litigious panic and misplaced health and safety anxiety could rob my daughter of that purest, simplest and most memorable of early childhood experiences – the simple act of playing in the snow with her friends. In the three-page letter that the head teacher received the following day I acknowledged that yes, there may be wet clothes, bruises, God forbid even a broken arm, but for my daughter and her friends to have this experience stolen from them was a crime beyond redemption. Suffice to say by return of post I received a three-line letter from the school quoting the relevant health and safety regulations. Out-*bloody*-rageous.

Not long after we visited friends in Sweden and spent some time at their children's school. It was winter and very snowy and icy and the children were all out in the playground with the teachers having a great time. This included a group of kids who had forged a perilous seven-metre ice slide on the concrete playground down which they would hurl themselves and each other at considerable speed, often crashing into the children who were gingerly picking themselves up at the far end, sending them flying in all directions – a kind of human ten-pin bowling. Now that's the sort of snow fun I'm talking about.

Speaking of Scandinavians hurling themselves around, my thoughts turn to my mother who is an incredible woman. Nurse, magistrate, Samaritan, Quaker, rampaging socialist, tireless charity worker... the list goes on and on. I recall that she spent much of the early eighties travelling down to Greenham Common, the site of the infamous RAF airbase, to protest against the government's decision to allow American cruise missiles to be sited here in the UK. Even now, in her early eighties, she devotes all her time to supporting people around her in all manner of ways. She is indeed the living embodiment of altruism; her whole life has been devoted to the betterment of those less able, less fortunate and less empowered. OK, so there are philosophical debates to be had about the nature of 'true' altruism – can such a notion truly exist in the context of the intrinsic rewards of personal gratification? The answer is clearly yes in my mother's case; her altruism is something that lies at the very core and essence of her being. In addition to everything else, she provided the centred counterpoint to my father's rather erratic, elliptic orbit around the family and without her grounding, nurturing presence our childhood may have been quite a different story.

She was (and still is) a wonderful cook and within the often freezing, crumbling and bohemian depths of our old house, various combinations of family and friends would invariably be gathered in the kitchen around the oily warmth of the Aga, eating warm home-made bread and split pea soup, trying to solve that tricky final clue in the *Guardian*'s quick crossword. One of our cats would occasionally join us, actually sitting in the lower oven of the Aga – a warm but risky option, though I thankfully can't recall any oven-ready cat episodes. My mother's motto, which she sought to pass on to us children, was always to 'live every day as if it were your last' and she has certainly lived up to that – packing more into a lifetime than many of us could ever hope for. She certainly climbed many mountains in her time; tests of strength, stamina and determination that I guess go right back to her own childhood days in Nazi-occupied Denmark.

It is interesting how our present lives are shaped by our childhoods, just like these mountains are shaped by the turbulent history of their own infancy. The child is the father of the man – the mother of the woman – and all that. Even now, my mother talks movingly about the memories imprinted upon her mind from early childhood; memories of her school being taken over by the Nazis and their houses being frequently searched (although not well enough for the Nazis to discover the stash of hand grenades that my uncle had hidden in his stove). We are but a collection of stories that we carry through time, sometimes fluid and coherent, sometimes grating and abrasive as they mix, slide and collide into one another like our very own set of personal tectonic plates, forged in the fires of our formative years. Hitler was also in the Alps of course, his presence never that far from either of my parents' past lives. He certainly left a large footprint in the snow in these parts; his home of Berghof at Obersalzburg in the Bavarian Alps being where he spent most of his time during World War Two, and the infamous Kehlsteinhaus – the Eagle's Nest – perched on the mountain top above Berghof, where he in fact went very rarely.

I have my mother to thank for many things, not least the direction that my life took after a long period of aimless drifting following a rather undistinguished school career. Caught between my mother's fervent socialism and my father's ambiguous but often extreme conservatism (the political differences perhaps symbolic of the marital differences that pushed them ever further away from each other), my education was something of a 'curate's egg' – good in parts. I have fond memories of my early school years; I attended the local village primary school that consisted of about 35 children in three classes with three teachers, but from here I moved to an enormous comprehensive school of close to 3,000 pupils. At the time it was the largest school in the southeast of England, and it really was a case of the sublime to the ridiculous. I recall our primary school head teacher giving us the routine big fish/little fish talk, but he didn't say anything about being thrown to the sharks. I trod water for a couple of years and then pretty much sank without trace, passing out with a paltry three O levels to my name.

There were many reasons why I crashed out at school, some of them best not mentioned here, but the main reason was that I just wasn't interested. The school was too big, too impersonal and I was miserable for most of the time. I remember my personal tutor taking me aside one day and saying, 'Listen David, just keep a low profile and you will be fine.' Admittedly, the fact that I had ridiculously long, curly hair that reached nearly to my waist meant that keeping a low profile was a little difficult. (Whatever you do, don't look conspicuous!). I took her words to heart and have followed her advice ever since. I didn't know quite what she meant at the time, but it's a good policy that has served me well over the years. Mrs Ellway, bless her. I think she saved me.

So I left school to join a band (rather than the circus), never went to university (let's face it, standards might have been lower back then but three O levels didn't quite cut the mustard) and spent a good number of years in the wilderness – the emotional Serengeti of my adolescence. Mind you, aimless drifting is not all bad, in fact it was mostly fun and the other thing that saved me, as well as Mrs Ellway, was music. We can track our progress through life in a number of ways: jobs, relationships, places we've lived, joys and tragedies, successes and failures, regrets and opportunities; just as Rupert and I track our progress around Mont Blanc. All of these experiences provide a part of the overall story, the narrative thread that weaves its precarious way through the years, binding and wrapping us together in a vibrant collection of experiences and stories that make us who we are. And one particular thread that has been a constant throughout my life has been that of music – music that I have played and the people that I have played alongside. For the last 35 years I have always played in one band or another, a beautiful river of noise that has meandered its way through the decades. And it really has been a meandering; there has never been a plan or indeed any great ambition. I simply enjoy the notes and sharing them with other people. Since leaving school at the age of 17 to my nearly-arrived-at half century, there has probably been little more than a matter of months when I have not been playing in a band of one kind or another. And I'm not choosy. Along the way I have played in punk, funk, pop, grunge, folk, experimental and jazz bands. You name it, I've played it and I have loved them all.

Fame and fortune have eluded me on the musical front, but then I have never especially hankered after fame and fortune (the very idea frightens me to death). The most successful of all these bands was a whacky outfit called The Weeds. We played a kind of psychedelic grunge – imagine a cross between The Doors and The Velvet Underground with a little sprinkling of the Sex Pistols thrown in; chaotic, messy, sometimes brilliant and often awful as we rode the heady wave of the late eighties indie scene. As one *New Musical Express* reviewer memorably wrote, 'dirty, unwashed and unwanted'. But we got regular reviews in the national music press, mostly good, toured the country supporting Primal Scream and would have played at Glastonbury had our drummer turned up. Ah, those were the days. These days I am enjoying playing in what has to be said is a rather good jazz/funk band... available for weddings, bar mitzvahs and funerals. So music has been a constant throughout my life, and as I said may have been a saviour in my troubled years as I struggled to navigate my way through the turbulent waters of teenage angst. But as I was saying, it was primarily due to my mother's implicit influence and intrinsic social values that I finally escaped the post-school fog of career inertia and drifted towards my present work as a child therapist, via training in social work, dramatherapy and play therapy. Thank you Mum.

But it was more a case of drifting rather than any kind of plan. In fact as a young child my greatest single obsession was with nature and wildlife, a direct result of a childhood that saw me deeply immersed in the Sussex countryside and fascinated in all things creepy-crawly. My heroes of the day were David Bellamy, David Attenborough, Jacques Cousteau and Gerald Durrell. Far from being a therapist, my own childhood fantasy was to be a naturalist, palaeontologist or zoologist of some sort. In fact one of my elder brothers stole my dream job as a marine biologist. Damn! How did that happen? Just like my younger brother stole my much-treasured memory of sitting on Jenny Agutter's knee when I was seven years old, but that's another story. Practically my whole childhood, from the age of five to eleven, was spent conducting intensive biology fieldwork around the rivers, lakes and fields that surrounded our house. I was mostly preoccupied – obsessed I should say – with the fish pond at the end of our garden. It was probably just a small, regular kind of pond, but for my five-year-old self it was a whole new universe to explore.

Even now, 45 years later, I can recall the warm, silent glow of the late afternoon sun and the weedy, rosemary smell of that pond that always teemed with life of some kind. There were the whirligigs; tiny black beetles that traced dizzy patterns upon the surface, using the tension that separated water and air as a bumper-car playground. The whirligigs patrolled the surface, only occasionally pausing for brief moments of respite. Then there were the pond skaters, also surface dwellers, who skittered in a more leisurely, circumspect fashion. And beneath the water, just below the surface, the water boatmen chugged, legs like little oars, heaving themselves through the warm, sun-heavy water. I always wondered about the relationship between these strange creatures: the whirligigs, pond skaters and boatmen. They were the sky dwellers of the pond, cruising the high water above the stillness far below.

Even now the fish pond sits in my memory, an oasis nurturing the less well-fed recollections of my childhood, a place of warm stillness to which I sometimes return, as I often did when a child. I recall every drop of water, every weathered flagstone corner, every crevice of time-worn stone and fibre of pungent, green moss. In fact I think I nearly died in the pond if my memory serves me right. These kinds of memories are partly mine and partly of others who have stronger powers of recollection; perhaps in this sense memory is shared, not owned. But I do have an image seared into my five-year-old mind, an organic imprint upon my then still forming neural pathways that spread and multiplied like the weed within the nutrient rich waters of the pond. It is of my father's face, distorted by water, hovering wraith-like above me as I helplessly drifted towards a bed upon which I was surely never meant to rest. And then being held, his hand in mine, and pulled firmly through the water to safety. Perhaps he saved my life? I misremembered once that he had actually been holding me down. Still waters run deep; stagnant pools beside a memory.

Eight

Having finished our hot chocolate and having no real justification for hanging around in the Refuge des Mottets any longer, much as we would like to, Rupert and I agree to press on and so heave our rucksacks onto our protesting backs and once again hit the path. The route from here heads north-eastwards and takes us away from the stream and up the side of the valley and while the going is not overly demanding it is fairly steep at times as the path hairpins its way upwards. The Refuge des Mottets is soon left behind and as we gain height we get a dramatic, sweeping view of the Vallée des Glaciers behind us. All in all it is about a 750-metre climb and the path, initially a well established and fairly even gravelled affair, soon breaks up into a tangle of little paths and trails as the slope broadens out ahead. Still, it is hard going and being only our third day of walking I think that we are both still in the process of finding our 'mountain legs'; those mythical appendages that long distance hikers so often aspire to.

About an hour and a half after leaving the refuge we reach the Col de la Seigne that stands at 2,516 metres and is marked by a large two and a half-metre-high cairn. This is the border between France and Italy and the views are breathtaking. Kev Reynolds, our trusty pocket guide, describes this moment as a 'revelation of a new world' and he is absolutely right. There is a stunning sense of a new realm opening up ahead of us as we look forwards into Italy and back into France and the views in every direction, wherever your eyes take you, are simply overwhelming. Ahead of us the hillside falls away into a vast, deep channel flanked by the mountains on either side. To our left, Mont Blanc maintains its majestic presence – a great chunk of whitened rock rising behind the saw-toothed pinnacles of the lower mountains. The good weather has continued although there is a stiff breeze and a biting chill in the air, but the two of us feel totally blown away by the magnificence of this place and just loiter for a while soaking it in, reluctant to move on despite the cold wind that whips down the valley.

There is a special, deeply evocative quality about these borderline places; the notion of movement, transition, history and as we take in the panoramic grandeur of the Col de la Seigne it conjures up a powerful spirit of the past. This would have been an ancient high crossing between Italy and France, used by the Romans and those that followed throughout the Middle Ages, a gateway to the great Aosta valley that is bordered by France to the west, Switzerland to the north and the Italian region of Piedmont to the south and east. A short way down the hill from the col on the Italian side is La Casermetta, a renovated grey-stoned former customs house that is now used as a museum and mountain environmental/ information centre. It is a mark of bygone days and going back further was in fact a small military barracks; a border outpost manned by garrison troops and it remains a poignant symbol of the sometimes difficult history between France and Italy as they rubbed shoulders up here in the Alpine highlands. As tension increased during the 1930s this area was used for high altitude military manoeuvres and during World War Two the Italian army launched attacks against the French, already defeated by the Germans. The remains of military fortifications and shooting posts still dot the area and all in all it is a sober reminder of at least one story of these remote border crossings.

We begin to leave behind the Col de la Seigne and head downwards and Rupert, deciding he needs an energy boost, magically produces a Twix bar from his rucksack. As he opens it a sudden gust of wind rips the wrapping from his hand and sends it swirling back over across the col. Perhaps because of my earlier hat recovery experiences I take matters in hand and slip off my rucksack and head off in hot pursuit. The Twix wrapper scoots along about three metres ahead of me and soon I am back in France again and at this rate it won't be long before I am back at the Refuge des Mottets. A number of other hikers whoop encouragingly, enjoying the thrill of the chase and just when I think I am beaten the pesky wrapper lodges against a rock allowing me to snap it up. Victorious and to a smattering of light applause I trudge up the hill and back into Italy where I find Rupert just finishing off his Twix. '*Grazia mille amico,*' he says, licking his lips. Betwixt and between – what better chocolate confectionary to symbolise the essence of the border crossing? Admittedly, as border skirmishes go, this episode is just a minor footnote in Franco/Italian history. I doubt somehow that it will become the stuff of legend, but who knows?

And so we begin our descent into Lée Blanche, the upper reaches of the Val Veni, and it is not long before we reach the pastures and grasslands of the lower levels. The track, once rocky, snowy and muddy opens out into a wider gravel path and the going is pleasant enough as we leave the cold behind and move into the comparative warmth of the lower valley, lined with its tumbling slopes of grey scree. Occasionally, the gently hypnotic and randomly percussive jangle of cow bells floats across the air along with the occasional whistle of the marmots that watch us pass by with a curious concern, signalling our progress to their furry friends down the line. As well as the ever present visual spectacle that surrounds us, I love the evocative aural soundscape of these mountains and the tantalising memories and associations of past times that it conjures. It is late afternoon and the blue sky has given way to banks of grey cloud that seem to have stealthily closed in and drawn a murky veil across the mountains on either side, even proffering a few drops of rain as the day begins to close in. We are both beginning to feel quite weary, so it is with some relief that we see the Rifugio Elisabetta ahead to our left, perched on a spur about 30 metres up and nestling just beneath the Glacier d'Estellette that reaches down from above with its icy tongue.

We climb up the slope to the refuge and make our way in to find that the place is heaving. The dining area is already packed with people resting, drinking and eating and there is a frenetic buzz of activity as tired hikers barrage the stressed looking staff with requests for a space to bunk down for the night. Apparently 10,000 people from all over the world walk the TMB each year and it feels like every single one of them is here tonight, in the great melting pot that is the Rifugio Elisabetta. People jostle good humouredly for a little bit of space to rest while others, not quite so good humouredly, are turned away – no room at the inn – and have to continue on, the only option at this late stage in the day being another two hours' walk in the gathering rain to the hamlet of La Visaille where there is a bus service to Courmayeur.

Having been warned that it might be an idea to book ahead, we had optimistically asked the kindly folks back at Bonhomme to phone through and reserve Rupert and myself a bed for the night, but we are not sure whether this message actually got through. The refuge manager gallantly holds court at one end of the bar, clutching 'The List' (more a scrawl of names that have been underlined, crossed out or highlighted with all manner of strange symbols) and deftly fends off the increasingly fractious demands of tired hikers who would do anything rather than have to pull on their boots and head back out into the settling light of the early evening for another two hours walk.

Eventually, the moment of truth arrives and while it initially looks hopeful that my name is on the aforementioned list, there then follows an anxious ten-minute debate about the precise spelling and pronunciation of 'Le Vay'. The manager grills me as to my true identity – Levin? Levitt? Leighton? Lichtenburg? Blimey, how many bloody 'L's are there here tonight? It would have been a hell of a lot simpler if I had just become John Smith for the night and I also don't want to mistakenly bump anyone else off the list and get involved in an unedifying showdown with someone called Daniel Levinstein or some such – a bizarre kind of name-off with the winner getting a bed for the night.

As it is, the manager and I eventually agree on what seems to be a rough, phonetic approximation of my name and with a relieving grunt of approval we get a little tick on the list and he tells one of the staff to show us where to sleep. Well, as Rupert and I are led through the tangled mass of hikers to our 'room' we begin to wonder whether the walk and bus to Courmayeur might have been a better option. The dormitory is a three-tiered platform affair, each tier holding anything from ten to fifteen people, each space delineated by a sleeping bag, rucksack or already recumbent body. A sliding ladder system moves between the tiers, like some kind of Rubik puzzle, and the guy simply gestures to an ill-defined 'space' on the top tier (already full) and indicates where we are to sleep. Hmm, I wonder if this is organised alphabetically, there being a good chance that I am going to be spooning with Daniel Levinstein tonight.

To manage the hordes of hungry hikers, the refuge puts on two sittings for dinner so while the first lot eat we try to sort ourselves out in the dormitory. There are clothes everywhere – damp socks, pants, shirts and hats hanging from any available surface, and semi-disgorged rucksacks line the floor around which we have to carefully navigate. There is a rather alarming degree of collective rummaging going on and in the confusion I am sure that once or twice I accidently have a quick rummage in someone else's sack, so to speak. The shower is operated by a token system, each token providing a luxurious 90 seconds of lukewarm water, and so we have to queue, half-naked, clutching soap, shampoo and toothpaste etcetera and share tactics as to the best way to approach the matter in hand. The extent to which one lathers up seems to be the question of the moment or, to be more precise, the relative dangers of over-lathering. This issue seems to have been provoked by the sight of one or two people who have clearly misjudged their 90 seconds, gone for the 'big one' only to emerge from the shower looking like some kind of soapy abominable snowman, perhaps down for an overnight excursion from his lair in the Glacier d'Estellette that looms above us. As for the toilets, well the less said the better, suffice to say that a few of us executed carefully timed raids on the toilets belonging to the dormitories upstairs, which were slightly more inviting – or less uninviting I should say.

Later we manage to eat, grateful for a space to sit down for a while, and we finally begin to allow ourselves to relax and accept the challenge of the night ahead, helped no doubt by a carafe or two of red wine. It is a kind of madness here at the Rifugio Elisabetta, but in the midst of madness there also lies a strong sense of camaraderie. We are, as a group, thrown together and it is undoubtedly a bonding experience. Our friend Richard the surveyor has rejoined us after making his way here from Les Chapieux, as have the two camping girls who we walked with for a while on our first day. We chat and joke together in the dormitory along with an international cast of other characters who mostly are all throwing themselves into the spirit of the occasion. A Dutch woman who is walking on her own and two charming South African women, probably in their mid-sixties, are great fun and most of the conversations centre on how on earth we are going to get through this night together. Some people prepare to sleep on tables, some on the floor, some squashed on benches below the windows and a number of deals, swaps and even bribes are negotiated for the best possible sleeping spots. The other issue that merits much discussion is that of snoring and with forty or so people crammed together in this one small dormitory, this is no trivial matter. Rupert and I come to an agreement that if either of us is snoring we can shove, kick or punch each other until we stop and with that agreed it is lights out and we all, one by one, climb, clamber and haul ourselves up and into position and drift off to the soothing warm fug of wet socks, methane and CO_2.

Nine

I slowly regain consciousness at about 6.45 a.m. and it has not been an easy night, but I am thankfully still alive. I remember having to give Rupert some sturdy shoves in the middle of the night to stop him snoring, and when I turn to see if he is awake or not I discover he has gone missing. Where can he be I wonder? Perhaps he got up early to sketch the sunrise, which if the case would impress me no end. Slowly the line of sleeping bags on either side of me are starting to come to life, a row of human cocoons preparing to hatch out – like a scene from the *Invasion of the Body Snatchers*. I clamber down the ladder and go in search of Rupert and discover that he has slept the night in the boot room, lying on the floor among a hundred pairs of smelly boots and socks, trekking poles casually scattered around him as if he has been playing a nightmare giant version of pick-up-sticks in his sleep.

To my own horror, I discover that it wasn't Rupert who was snoring but the person next to him; poor Rupert got so fed up with me prodding and shoving him that he got up in the middle of the night and decamped to Boot Central. I feel mortified at my mistake and apologise profusely to Rupert who is remarkably gallant about the whole episode. To my further horror, at breakfast Richard tells me that actually I was snoring rather loudly myself during the night (apparently 7/10 on our snoring Richter scale). So I am doubly at fault; it should have been me in the boot room. We spend most of breakfast, along with the whole of our table, going through the post-night analysis of my misdeeds and Rupert's misfortune. Like wildfire, the story goes around the refuge and it seems to be the only thing that people are talking about: 'Did you hear about that guy who drove his friend out to the boot room because he thought he was snoring – even though he wasn't – and guess what? He was snoring himself.' Oh well.

Overall it has been quite an experience at Rifugio Elisabetta and certainly something of a TMB bonding session, but it is with some relief when we head off at about 8.00 a.m. To be fair, it's a great refuge in a superb position; it was simply the amount of people that contributed to our rather unique experience. Certainly, I would come here again, but would perhaps try to pick my time with a little more care. Richard is with us, but the 'camping girls' – who we have discovered are in fact named Julie and Claire – have left ahead of us.

The first stage of the walk is very pleasant, an undemanding stroll along the valley floor as we follow the course of the river on a small, level track that provides a degree of respite from the ups and downs of yesterday's walking. After about 45 minutes we come to Lac Combal, its milky-grey water flanked above by a perfect, almost man-made looking wall of lateral moraine, the parallel ridges of debris that have been deposited along the side of the glacier. And then there are the glaciers themselves, no doubt shades of their former glory but majestic all the same as they lie embedded in the folds of mountainside – solid, ancient and almost unearthly in appearance. The sound of meltwater, tumbling and trickling its way into the stream that runs along the valley floor and the ever present melodious toll of the cow bells, like living wind chimes, combine to captivate the senses.

It is a scene that is at once beautiful, wild, rugged and serene and it is easy to see how this 'new world' captured the fevered imagination of the writers, poets, and explorers of the eighteenth century. As our aforementioned friends Balmat, Paccard and Saussure *et al.* paved the way to the summit of Mont Blanc, forging a path for others to follow, there was also during this period something of a cultural sea change at play that transformed the way in which these mountain landscapes were perceived.

Central to this perceptual shift in the middle of the eighteenth century was the formation of the Concept of the Sublime, a cultural and intellectual construct that fundamentally altered people's relationship with the natural environment. More than that, it was a doctrine that revolutionised the relationship with landscapes alongside attitudes of the time around feelings of fear, excitement and, to an extent, chaos. It was certainly a doctrine that challenged the prevailing neoclassic aesthetic of the day, and was a philosophy that delighted in the rugged, the fierce, the wild and the dangerous, and it was the mountain landscape (and most particularly the Alpine landscape, which was coincidentally becoming more accessible), once shunned and feared that truly encapsulated this brave new world.

The foremost proponent and chronicler of this new way of thinking was the Irish writer Edmund Burke whose 1757 work, *A Philosophical Enquiry into the Origin of Our Ideas of the Sublime and the Beautiful,* sought to consider the passions and emotions evoked in the human mind by what Burke termed 'terrible objects'. He was interested in our psychological response to experiences that both excited and terrified; experiences that through their size, complexity, danger and potential uncontrollability evoked a visceral and tantalising blend of fear and pleasure. At the heart of Burke's thesis was the notion that these sublime sights and experiences caused a degree of terror, and terror was a passion that 'always produces delight when it does not press too close'. Indeed, as Burke writes, '... whatever is in any sort terrible, or is conversant about terrible objects, or operates in a manner analogous to terror, is a source of the sublime; that is, it is productive of the strongest emotion which the mind is capable of feeling'.

So, it was the suggestion of harm, the proximity to danger associated with the knowledge that no actual harm was likely to result, which induced this exciting frisson of sublime terror. In a way, this relates back to the idea of children and risk, the capacity to be able to safely experience a degree of danger – the essence of which is so much part of childhood, as I was saying earlier. As a therapist, I find this notion of the 'sublime' and the desire to confront something that both terrifies and excites, fascinating. There is a sense of these unknowable, uncontrollable external landscapes, the ravines, crevasses and rugged peaks of the mountains, reflecting something of our own internal landscapes – the fears, fantasies and dangerous dark places that to an extent inhabit all of us and this seemed to be something of a subtext to Burke's philosophy.

There are also perhaps echoes here with what we therapists might think of as the 'Jungian shadow' – that part of ourselves that we least like to imagine and of which we are not fully conscious, or as Burke might put it, 'the dark, confused, uncertain images' of the imagination. This is the dark psychology of the mountains, symbolic of a primal unconscious, ancient and archetypal as Jung might see it, and the mountains provide something of a mirror to the highs and the lows, the peaks and troughs of the human spirit. Thinking again of the small vial of ashes tucked away in my rucksack, there is a uniquely bipolar quality to this landscape that perfectly captures the spirit of my father and I understand perfectly why he so liked these places himself – a soul reflected. Carl Jung was Swiss of course, himself also subject to the influence of the mountains and I wonder about the part they played in the formation of his own personal psychology.

As well as Edmund Burke, the other figure that played a key part in this reinvention (or perhaps reimagining) of the mountains was that of Jean-Jacques Rousseau, an eighteenth century writer, philosopher, composer, social theorist and leading intellectual figure of the day. Rousseau is counted as one of the prime movers in the development of the Romantic movement in art and literature and predecessor to the likes of Shelley and Byron, themselves so entranced with the Alps in later times, and he is seen by many as playing a pivotal role in shaping and changing our ideas about the mountains and specifically the Alps. In his novel *La Nouvelle Héloïse*,[9] which was read widely throughout Europe, Rousseau paints a picture of the Alps as a place of serene tranquillity and benign purity – a stark contrast to earlier writers who described the mountains as dark places of torment and evil. His central character, Saint-Preux, seeks calm and solace by going for long walks in the Alps. And in his autobiography, *The Confessions*, Rousseau writes, '... there is a kind of supernatural beauty in these mountainous prospects which charms both the senses and the minds into a forgetfulness of oneself and everything in the world'.[10]

So whether a result of the Sublime or the Romantic, most likely a heady concoction of the two, Mont Blanc and the Alps find themselves transformed, reimagined and reinvented through a process of cultural alchemy from a place of fear and dread to a tantalising new world full of hope and promise. Certainly, as Rupert and I walk in quiet reverie past the misty waters of Lac Combal, the jagged dark-grey peak of the Aiguille Noire de Peuterey reaching into the blue sky ahead of us, it is hard to think of a better and more astute description than Rousseau's of the qualities of this place. One does indeed 'forget oneself and everything in the world', so entrancing is the landscape and it is easy to see how the work of Rousseau and others of the period provided a ground-breaking reimagining of this great mountain range.

At the end of the lake the path turns right and angles up the slopes of l'Arp Vielle and, while on paper it looks like a reasonably routine 460-metre or so climb to the top, it's actually pretty hard going as the steady gradient requires a bit of effort to push ever upwards in the increasing heat of the morning. The track itself is well worn but rocky in places and demands concentration as it twists and turns its way up the steep side of the valley. As we gain height the southern face of the Mont Blanc range is dramatically revealed, increasingly stunning the higher we climb, and directly across the Val Veni is the Glacier du Miage that sweeps down in a great dirty river of ice and rocky debris. At close to ten kilometres in length this is the longest glacier in Italy and at least half of the glacier is covered in grey debris, mostly the result of rockfalls from the surrounding walls. It is a striking sight and as we climb higher we get a spectacular sense of the immense scale of this great mountain range, more so than at any other time over the last few days. There is clearly some considerable distance from where we stand to this southern flank of the Mont Blanc range that rises up like a wall on the other side of the valley, but it feels close enough to reach out and touch, the clear morning air adding to the tangible quality of the scene before us.

Last night's clouds have been blown away and once again it is turning out to be a beautiful day, hot and sunny, and we can't believe how lucky we are to have such fine weather. Richard is good company and regales us with tales of previous walking adventures and it is not too long before we catch up with Julie and Claire, who are ambling along at a leisurely pace. And so we hook up as a band of five and merrily make our way along the trail, exchanging stories, jokes and anecdotes as we go, a little like a scene from *The*

Canterbury Tales, the five pilgrims en route to Courmayeur – although I doubt there will be a prize of a free meal at the end of the day.

The route is busy and we are surprised by the line of hikers that dot the trail both ahead and behind us and perhaps, like us, many people waited until the TMB ultra marathon was over before starting the walk, resulting in large groups setting off at the same time. But busy as it is, there is a lovely sense of camaraderie as we occasionally stop and chat to little clusters of fellow hikers and exchange stories and news and useful snippets of information. The buzz on the TMB grapevine right now is all about last night at Rifugio Elisabetta and everyone has their personal story about how they survived the experience. Now and again, as we chat to people we meet along the way, someone might say 'Ah, so you are the guy who kicked his mate out of bed and made him sleep with all the boots. Yeah, we've heard about you.' As fame goes, it's rather inglorious.

It turns out that Julie and Claire both work for the RSPB in Scotland and are fun walking companions. They have a quirky sense of humour and provide an amusing running commentary to our progress on the trail. They also clearly share a deep affection for the natural world and are incredibly knowledgeable about the local flora and fauna, frequently stopping to take pictures of insects and tiny alpine flowers with the macro settings of their cameras. It is a little like being on a guided nature walk as every now again they might pause and point out a particular flower, the lilac bell-shaped campanula perhaps or stop to listen to the high-pitched chirrup of an alpine chough. Like Rupert's sketching, these moments serve to slow us down to really absorb the beauty of the region as we pass. It is not too long before we hit the high point of the day, sitting at 2,430 metres in all its panoramic alpine glory, and from here we drop down on a wonderful path that takes us into the beginnings of a pine forest. The sun is warm on our backs, the heat mixing with the rich, fresh alpine smell of the trees and butterflies lace the perfumed air in a colourful abundance. On the ground hundreds of tiny grasshoppers precariously leap ahead of our steps, revealing vibrant blue and red inner wings that are only visible when airborne. This is about as perfect as it gets and we can't but help walk with broad smiles on our faces as we share in the simple joy of the moment.

Rupert is particularly pleased with his merino base layer and has taken to mentioning it at least ten times a day and even stops at one point to wash it in a stream, which means he can talk about it even more to the passing hikers who pause to watch him with interest. So much for not washing your dirty linen in public. After a brief group consultation we ban Rupert from mentioning his undergarments more than three times a day. His blister has well and truly calmed down thanks to the Compeed and he is in good walking spirits, although he is perhaps, on this fourth day of our walk, just beginning to question whether he might actually have packed too much stuff in his rucksack.

We stop for a panini lunch at the Rifugio la Maison Vielle on the Col Chécrouit and it is a lovely spot, with views in one direction across the valley towards our constant companion of the day, the Aiguille Noire de Peuterey and southwards along the Val d'Aosta towards the distant fringes of the Gran

Paradiso range. The place is buzzing with day trippers from Courmayeur and it is clear that the Maison Vielle does a brisk trade in summer and also in winter when the skiing season kicks in. The sun is scorching and we have to seek the safety of some shade while I indulge in a refreshing Coke and Rupert, still in merino heaven, goes for a cheeky beer.

Ten

Rupert is a fine friend. We go back almost 17 years to when we met on our first day in a new job that we happened to be starting together. We were both social workers back then, untarnished by the cumulative cynicism, stress and organisational fatigue that tends to set in after a given amount of years on the front line. The most stressful thing about social work is working within an organisation that can't decide what or how it wants to be. Over the years I have been restructured, reorganised, rationalised, redefined and reshaped more times than I care to remember. I have been joined up, benchmarked and drilled down so often that I wondered if I had accidently drifted into a career in carpentry. Management jargon is, of course, more about obscuring the truth than clarifying it, a kind of corporate strategy to depersonalise so that the real impact of change is diluted through a filter of nonsense; dangerous, insidious nonsense some might think. So my once youthful rose-tinted specs have been seriously dulled over the years. I speak more for myself here; Rupert has always been better able to hold on to a degree of optimism. I generally maintain a very cynical world view and even have my very own mug with the words 'half-empty' printed on the side in big red letters. Nicky (who has her own 'half-full' mug) would probably say that mine isn't even half-empty, it's practically bone dry such are the heights – or depths – that my cynicism can reach. But I maintain it is a good position to hold; it saves disappointment and is a place from where things can only get better. Anyway, 1 April was the day when Rupert and I started in our new job (an appropriate day some may think – a portent of future foolish endeavours) and we cemented our friendship by decamping to the local pub at lunchtime to compare induction programs. In later years, both Rupert and I left social work behind to train as therapists – he as a family therapist and myself as a play therapist – and so our chosen paths, while divergent, have taken us over similar territory.

As I say, he's a good friend, kind and generous of spirit, although he did accidently once set a dog on fire. OK, this may be a little harsh. It was actually back in Cornwall on New Year's Eve, the very same time that the plan for this walk around Mont Blanc was hatched, and it's fair to say that it was a rather ominous beginning to the New Year. We were standing on the breezy beach a few minutes after midnight, glasses of sparkling wine in hand when Toby – a small fluffy dog of indeterminate breed – ran smoking across the stones like some kind of canine version of the Red Arrows air display team. Horrified and dumbstruck, everyone froze for a moment and 2012, just the merest of twinklings in the eye of Old Father Time, came to a temporary standstill. We had not anticipated that as we saw the New Year in with a few lightweight fireworks and a midnight dip in the Helford estuary, a small dog would suddenly appear out of nowhere and do an excited impersonation of the phoenix rising from the flames. And it was Rupert's firework (a Lemon Bombard, I believe) that especially caught poor Toby's attention. (Well actually it caught a little more than just his attention.) Thankfully, Toby was fine and was left with just a slight singeing to the underbelly and was otherwise unwounded apart from his little doggy pride.

Toby's owners were understandably not best pleased and although blameless we had a tricky few moments on the beach as we attempted to ascertain exactly what had happened. It was quite a traumatic experience and neither Rupert nor I can go anywhere near a firework now without uncomfortable flashbacks to that episode – and I guess the same might be said for Toby. Rupert was no more responsible than anyone else, apart from the fact that it was his Lemon Bombard, but when we are out socially it is hard to resist the temptation to introduce him as the man who accidentally set fire to a dog.

We have a dog ourselves, a black lab called Sadie, whose arrival as a puppy was payback (possibly even a replacement in case I fell off the top of Mount Canigou) for me disappearing to the Pyrenees for two months some years ago. Not being a dog person it took the two of us – Sadie and I – a while to bond but now we have a good relationship built upon mutual respect and Sadie's informed understanding that I am the leader of our family pack, however much she would like to be. Dog owners (and I still don't quite include myself in this group) are a curious bunch. Often, when we are out walking, someone will come up and say hello to Sadie and ask her things like whether she would like a biscuit or whether she is having a good walk. Now, being a dog Sadie of course can't actually talk so Nicky, in what she sees as her best approximation of Sadie's voice (imagine a high pitched, girlish voice with a touch of Frankie Howerd thrown in) will answer for her: 'Ooh, yes missus, I've had a lovely walk thank you. Don't titter. Ooh yes, I would love a biscuit. Oh madam please!' And so it goes, back and forth. Is this normal? Sane, rational people conducting lengthy conversations through their dogs in dog voices? Many a time I have been tempted to stop and ask a King Charles spaniel what they think about Nick Clegg's about-turn on tuition fees or whether spontaneous combustion was an effective narrative device in Dickens' *Bleak House*.

Fortified and refuelled from our lunch stop we depart the Rifugio la Maison Vielle and head downwards on a dirt track that passes by a ski lift, the cables disappearing through the trees and into the valley below. In the baking heat of the sun it is hard to imagine this place as a bustling ski resort and the various ski-related apparatus strewn across the area seems somewhat incongruous. In certain places the valley sides are fortified with avalanche barriers of wire mesh, netting and fencing – more indicators of the dramatic transformation of this landscape during the winter months. The path descends steeply and twists and turns its way through the wooded slopes of the valley. The lower we get the higher the temperature rises, inversely proportional, and the thinner air of the higher reaches becomes tangibly thicker, engulfing us in a blanket of heat and sweat that makes the going very uncomfortable. These steep descents are exhausting and the switchback path seems to go on forever, forcing us to stop and rest every now and again in a welcome bit of shade just to get our breath back.

Eventually, we break through the trees and hit a tarmac lane that leads us into Dolonne, the rather charming 'sister' town of Courmayeur, which lies west of the river that splits the two. We make our way through the narrow streets with their attractive, stone built houses and then have a tiring but mercifully short road walk that takes us across the river and into Courmayeur itself. We are hot, smelly and weary, but buoyed in the knowledge that tomorrow is a rest day. Ah, bliss.

Eleven

Courmayeur sits at the foot of Mont Blanc, like its French neighbour Chamonix on the other side of the Aiguille du Midi, with whom it also shares access to the renowned ski run of the Vallée Blanche. This is perhaps the most famous off-piste ski run in the world, accessed by the Aiguille du Midi cable car – the highest in Europe – that climbs more than 2,700 metres in around 20 minutes. On a good year a skier can get a run of 22 kilometres, sometimes more, with a vertical descent of more than 2,000 metres that includes the Mer de Glace, one of Europe's longest glaciers.

People frequently die on the Vallée Blanche, not necessarily on the run itself; the initial climb down from the cable car to the start of the run is by all accounts pretty precarious with skiers having to be roped up to prevent them plunging 1,500 metres to the valley floor. I am not a skier myself, but skiing 20 kilometres plus down a massive glacier sounds like my kind of fun. Courmayeur also shares administrative responsibility for Mont Blanc with the neighbouring French commune of Saint-Gervais-les-Bains. This essentially means managing the devolved political and economic responsibility of the 'commune' – at a local government level – hence allowing Courmayeur to lay claim to the title of the highest commune in Italy. (Although, unfortunately for Courmayeur the actual summit of Mont Blanc lies within the boundary of Saint-Gervais – allowing it sole rights to the highest point in Western Europe.)

Anyway, with the help of the kindly folks at the tourist office, we manage to book ourselves into the Pensione Venezia, a one-star backpacker hotel, which after last night's experience at Elisabetta feels more like five stars. It's a pleasant enough building, white with brown shutters and a rickety balcony that I nearly fall through as I lean against it to admire the view. Oh well, we're only three floors up. Behind the pensione, at the northern end of the town, the grey, craggy spires of the mountains rise up, dwarfing the buildings below. We get a double room for €50, with our very own beds and a shower just along the corridor and as far as we are concerned this feels like heaven. Richard, Julie and Claire are also in residence and it is a joy just to have a bit of personal space to spread out and collapse, wash, doze, write our journals and all the things that are so hard to do in the overly close conditions of the refuge bunkhouses. Neither of us can be bothered to hand wash our clothes so we go in search of a launderette, but the only one we can find is a place where you have to leave your clothes to be washed, dried and folded for you at considerable expense. It is late in the day and we are both tired so we agree, in the slightly uncomfortable knowledge that this is a very middle-class form of backpacking in which we are indulging. Is this just one small step from a personal valet?

Ever since we crossed the border back at Col de la Seigne, I have had to keep reminding myself that we are in Italy, but as we stroll around Courmayeur it suddenly feels like we have arrived. Although only a small town, it is smart and well heeled with expensive looking shops lining the main pedestrian thoroughfare, off which little alleyways provide some interesting nooks and crannies to explore. It feels classy and in that sense I guess, Italian. It is certainly laid-back and as we wander around in the late afternoon sun, enjoying the novelty of not being weighed down by our rucksacks, we bump into several of our fellow TMBers and still the main topic of conversation between us is last night's experience at the legendary Rifugio Elisabetta and the various sleeping heroics (or un-heroics in my case) that took place. So bonded by the experience have we become that a group of around ten of us arrange to meet up later for a drink (perhaps feeling more comfortable crammed tightly together in a small bar). I come to the conclusion that this is more of a trauma bond, lives forever bound together by an intense experience of inescapable threat. Will we meet up every year I wonder, and relive that long night of hell? Will Rupert spend years in therapy talking about how he was forced by his 'friend' to sleep in a room full of stinking boots? Will he have flashbacks every time he sees a trekking pole?

As it is, we meet together for a drink and while the rest of the group decide to go off and find a restaurant, Rupert and I have had a little too much group bonding and break away to go for a pizza in a great place called the Pizzeria du Tunnel. (The name doesn't really do it justice.) This is my first experience of eating a pizza in Italy and I am not disappointed. The restaurant is buzzing with both tourists and locals and the food is fantastic, as are the rather large quantities of red wine that we manage to get through.

I sleep very well and wake up about 7.00 a.m., slowly drifting into consciousness to the beautiful sound of church bells – the deep, sonorous ring of the bells creating something of a Doppler effect as the sound ricochets around the mountains. Having grown up close to a church the sound of church bells is hardwired into the nostalgia circuits of my brain, instantly transporting me to my childhood. Church bells would be an easy pick for my top ten list of best sounds, along with a cackling chorus of rooks, the *Shipping Forecast*, *Test Match Special* and the rambunctious farting of a Morris Minor 1000.

At breakfast, Richard, Rupert and I become engrossed in a long conversation about risk-taking when travelling and we share stories about foolhardy near-death experiences and how, all now being older and wiser, we would never put ourselves in these kinds of ridiculous situations again. Outside, Courmayeur is waking up and after a while Rupert and I go for a stroll around town. Fellow hikers drift up and down the main street; rucksack, boots and walking poles pretty much the standard uniform around here in the summer months. To emphasise the point, the street is lined with hiking, climbing and ski shops. A little tourist train, harsh in its crass appeal, occasionally disturbs the scene as it trundles on fat wheels around the town, it's annoying little bell a far cry from my nostalgic wake-up call this morning. Trains just like this one seem to be fixtures these days in places like this and they are an affront to a town's historic pride. They should be banned.

A large church sits high in the main square, opposite the Società delle Guide, an attractive traditional building that houses the Alpine Guides Society of Courmayeur, along with the Alpine Museum that was restored to celebrate the one hundred and fiftieth anniversary of the society. Originally built at the request of the Duke of Abruzzi, the prolific mountaineer, explorer and member of the House of Savoy, the building hosts a fascinating 200-year historical collection of material, including the written logs of deceased guides, in which their 'clients' have recorded impressions of their ascent alongside ratings for their guides' performance. Old photographs convey the steely strength, determination and resilience of these extraordinary men and an evocative collection of ice picks and boots displayed on the wall bring their brave exploits vividly to life.

Courmayeur can certainly lay claim to some of the most famous guides of Mont Blanc and beyond. These guides are in many ways the aristocracy of mountaineering and its traditions of the last two centuries. Evidence of this can be seen outside in the town's main piazza where there is a striking statue of Emile Rey, the 'Prince of Guides'. Rey was a member of the team that, in 1885, made the first ascent of the Aiguille Blanche de Peuterey, an ascent that is considered by some to be perhaps the greatest feat of climbing in the nineteenth century due to its challenging level of difficulty, height of over 4,000 metres and the fact that it had never been climbed before.

Courmayeur is also home of the Giardino Botanico Alpino Saussurea, an alpine botanical garden located 2,173 metres above sea level at Pavillon du Mont Fréty, the first station of the Mont Blanc cable car. This is apparently Europe's highest botanical garden and is situated on an elevated ridge south of Mont Blanc, and features rockeries, alpine pasture, scree slopes and wetlands. Opened in 1987, the garden is named after the perennial herbaceous plant, *Saussurea alpina*, with its blue and purple flowers, which in turn is named after our good friend Horace-Bénédict de Saussure. That man gets everywhere. The garden disappears under about three metres of snow for a good eight months of the year so if you want to visit, pick your time well.

Later we bump into Richard, who with a rather impish smile on his face tells us he has just got back from taking a 3,500-metre trip in a cable car up to Pointe Helbronner, one of the many peaks in the Mont Blanc massif. He looks impish because he spent all day yesterday telling people that the cable car wasn't running as he had been mistakenly informed that it had been closed for maintenance and was out of operation. It also turns out that, being a spur of the moment decision, he couldn't be bothered to go back to the hotel to change so went skittering about on a glacier at close to 4,000 metres in just his shorts, T-shirt and sandals. Hmm, what was that about risk-taking we were talking about earlier?

Rupert and I decide to go our separate ways for a couple of hours to give ourselves a bit of time out and do a few practical things like writing postcards. I end up doing nothing apart from sitting on a bench in some public gardens at the far end of town, simply enjoying the peace and quiet. From where I sit Mont Blanc, or Mont Bianco as I should say now, rises up to the west. The grey, rippled folds of the surrounding mountains form an imposing wall around the town, the lower sections dotted with patches of green pine. Against the blue sky, white clouds drift among the peaks, casting shadows that move gently across darkening patches of rock.

As I sit writing my journal, lost in my own little world, someone shouts a greeting and I look up to see a guy who I have talked with off and on along the trail – in fact I was wedged in between him and Rupert back at Elisabetta, so you could say we have slept together. He is a gregarious Belgian South African, sporting a generous white moustache that gives him the prestigious appearance of a character from a Jules Verne novel – a sort of old school, classic adventurer look. He is with his wife, who has flown all the way over from South Africa to meet him. We chat for a while about the walking and then he tells me that the woman who was sleeping next to him, back at Elisabetta, said that I was snoring quite loudly. This makes me feel even worse about kicking Rupert out under false pretences, although I am starting to think now that this story has really run its course and the TMB community should find something better to talk about.

I catch up with Rupert and we sit and have a bite of lunch in a cafe close to the tourist office. Over coffee and toasted sandwiches we watch today's wave of TMBers make their weary way into town, disgorged unceremoniously by the unforgiving mountains, much as we were yesterday. I inform Rupert, much to his chagrin, that my hat has been described by Claire and Julie as 'funky' and that after my daring act of hat retrieval the other day the said hat has reached close to legendary status on the TMB. I am also the only person on the trail using a traditional wooden walking stick and realise that among the rather brand-obsessed, tooled-up hiking community I must cut a rather eccentric figure. Some of these people take their hiking far too seriously with their camel bags, hi-tech gear and Velcro knee supports. (Velcro... it's such a rip-off.) Already I have been party to several conversations between people comparing the relative shock absorbing capacity of their walking poles. And there is far too much Lycra on show for my liking. This is just a long walk for God's sake; let's have a bit of common sense.

I also realise how like my father I am becoming, which may or may not be a good thing. But there is something about this walk, as when I walked the GR10, which takes me closer to his spirit and to the warm childhood memories that I treasure so much – nuggets among the more difficult memories from my teenage years when he became a colder, more distant figure. I tell Rupert about my father's ashes, stashed away in my rucksack. He is a little surprised at my disclosure, but curious also and interested in the meaning this has for me.

Being both therapists, we are pretty open to reflecting together about aspects of our own personal process: what makes us tick, what drives us, our respective vulnerabilities and what we euphemistically refer to as 'bottom drawer' stuff. This is the place where we keep the slightly dodgy stuff, the stuff that men perhaps rarely talk about; the feelings, fears, fantasies that we all experience but sometimes feel unable to share. We share stories about families, childhood

– experiences good and not so good – and our friendship is such that we can share these things safely, without judgment. True friendship should be unconditional; it needs to be able to tolerate the good and the bad, the strength and the vulnerability. There is ease in the relationship between the two of us, an ease that lends itself to a good walking companionship, and this is important when spending long periods on the trail with each other.

Saying that, I have tested Rupert's friendship to the limit by bombarding him with my repertoire of bad jokes over the last few of days. He has gracefully tolerated them, sometimes laughed but more often than not groaned. I like jokes, the worse and poorer taste the better. When I walked the GR10 in the Pyrenees, my friend Rob had to put up with my jokes for two months, off and on, and survived by honing his skills of selective hearing. When I was a child, my father used to bombard us with a constant barrage of bad jokes as we sat around the dinner table – mostly Jewish, often filthy – and he had a tendency to repeat them over and again over the years as if driven by some kind of obsessive/compulsive Tourette-like condition. Each time he would laugh uproariously while we would... well... groan. Is it some kind of genetic condition? I did a similar thing to Jess when she was growing up, although not quite as extreme I hope. Strangely, there is actually a condition called Witzelsucht, a rare set of neurological symptoms manifested through compulsive punning and joke telling in inappropriate social situations. I don't think I am quite that far gone, but Jess might say that sometimes it feels like a close run thing. Anyway, did you hear about the old Jewish guy who got knocked down by a car? The paramedic asks: 'Are you comfortable?' The man shrugs and says, 'Well, you know... I make a living.'

Rupert discovers that in the cafe they sell rather charming little furry marmot key rings and he rather touchingly buys us one each and also one for Jess, as a present for when we get home. We then think that Claire and Julie might like one and go back in and buy a couple more, finishing off their stock. They have probably never had quite such a run on marmot key rings. After idling away a couple more hours we return to the hotel and join Richard, Claire and Julie, who are sitting on the balcony of their room brewing up coffee on their little camping stove. Claire and Julie are admirably self-sufficient, having a limited budget for the trip, but by camping, cooking and generally fending for themselves they seem to be having a whale of a time. The girls are overjoyed with their marmots, which are ceremoniously attached to the back of their rucksacks. We didn't think to get Richard one, and he seems a little put out.

In the evening we all return to the Pizzeria du Tunnel, but I forgo the pizza in favour of a wonderful spaghetti carbonara. We have a fine evening, good food washed down with the obligatory carafe of red wine and all in all it has been a good day. It is interesting how these walks inevitably throw you together with other people, all on our respective journeys both literal and metaphorical. Richard is a lovely chap and he has a marvellous skill of being able to drift seamlessly between different groups of walkers, networking furiously as he goes. He is a little like a bee, buzzing busily between the fragrant clusters of hikers that he meets along the way, pollinating as he goes with his little snippets of information and advice about who is doing what, the options for the best route and the possible pitfalls that might lay ahead. He seems to know everyone on the TMB, and so we are honoured to have his company tonight. Warmed and sated by the food and wine, comfortable in each other's company, it feels good to be sitting in this little place, in this little town that nestles in the ancient glacial fabric of this great mountain range.

Twelve

Wrapped as we are in the timeless embrace of these icy mountains it is hard not to be touched by a sense of history, of the many people who have travelled these parts and beaten a rugged path to Mont Blanc's door, so to speak, in their haste to experience the new wonders that had been opened up across Europe.

One influential person to climb Mont Blanc was a young Scot by the name of John Richardson Auldjo, a graduate of Trinity College, Cambridge. A writer, geologist and artist Auldjo, like so many of the more wealthy classes in the eighteenth and nineteenth centuries, decided to embark on his Grand Tour in 1827 and the written account of his travels influenced and inspired many others to climb Mont Blanc. The Grand Tour was in its day something of an educational rite of passage in which, as *The New York Times* once put it: '... wealthy young Englishmen began taking a post-Oxbridge trek through France and Italy in search of art, culture and the roots of western civilization'. These were generally well-connected young people, often with aristocratic associations, who had the luxury of both the funds and the time to gallivant around Western Europe indulging their social curiosity, perfecting their language skills and generally mixing it with the cultural elite. With my own daughter Jess currently at Cambridge (crikey yes, they take ladies now) and knocking up a massive debt that will take half her lifetime to pay back, I am glad that this particular tradition has come to an end. Although perhaps it hasn't for many of the more well heeled.

It's interesting though. In the eighteenth century the control and influence of the ruling classes was to a degree located much more in a sense of cultural authority and supremacy than it was in expressions of military or economic power (as it is now). And so the Grand Tour served its purpose, introducing the younger generation to the value of art and music and the legacy of the Renaissance and reinforcing the cultural hegemony of the wealthy classes. It makes one wonder about the place of the arts within the current scheme of things; certainly there has been something of a turnaround in recent times. As Oscar Wilde famously said, 'Nowadays people know the price of everything, and the value of nothing.' But let's not talk about the bankers now. The advent of rail and steam soon made journeying around Europe much less of a burden upon money and time and hence more accessible to the middle classes, the way being paved by a certain travel entrepreneur, Mr Thomas Cook, the doyen of travel agents.

In 1827 and during the early stages of his Grand Tour, Auldjo decided that he wanted to climb to the top of the highest peak in Western Europe, as you do. When he arrived in Chamonix, Auldjo was sought out and warned not to make the ascent by a local man who had been badly injured in his own attempt. Not to be deterred Auldjo hired a small group of six guides and managed to make the ascent successfully. However, he suffered greatly in the process and it was mostly down to the concerted efforts of his guides that he reached the summit and survived at all. On the way up he suffered from altitude sickness, hypothermia and snow-blindness and on the way down heatstroke, intense stomach pain and near total collapse. It was said that when Auldjo was at his worst on the mountainside, suffering severe hypothermia and unable to continue, his six guides huddled around him and used their own body warmth to warm him up sufficiently to survive the final few hours of the descent.

Auldjo returned a hero to Chamonix, although as in so many of these accounts, it is the guides who are the unsung true heroes. When he eventually returned to England, Auldjo wrote up his experience of the ascent and it was the success of this narrative account (that ran to three printed editions) that caught the public imagination and inspired many other people to follow his rather tortured footsteps up the mountain. Indeed, as Auldjo said, '... perhaps it will not be decreed presumptuous in me to say that this brief narrative may be consulted with advantage by all those, who influenced by a congenial spirit of adventure, may be disposed to engage in a similar undertaking'.[11] Well, it certainly was and in the following years there was a steep rise in the number of summit attempts upon Mont Blanc by the British.

One of the people inspired by Auldjo's account and subsequently 'disposed to engage in a similar undertaking' was a certain young chap by the name of Albert Smith and it was the exploits of this man in the mid-nineteenth century, perhaps more than any other, that really fired the public imagination and possibly led Mont Blanc to become, for a period anyway, the most famous mountain in the world.

Albert Smith was born in Chertsey, funnily enough just a stone's throw up the M25 from where I live myself in southwest London. He was the son of a surgeon (now there's another funny thing) but I think that's where the similarities end. By all accounts Smith seemed something of a charismatic, flamboyant character – writer, satirist, theatrical impresario and general all-round showman. It would seem that the seeds of his energetic, eclectic career were sown at an early age when, at the age of nine, he became somewhat entranced with the Alps after reading a children's book called *The Peasants of Chamouni*.[12] Apparently, the young Smith would build small, moving mountain panoramas, like little stage sets, and entertain his family with frightening stories of climbers falling to their gruesome deaths on the perilous slopes. Perhaps driven by his early childhood imaginings, Smith actually managed many years later to visit Chamonix itself, in 1838, while he was studying medicine in Paris.

Although singularly unimpressed with the Mer de Glace (no more than a 'magnified ploughed field' according to his diary), Smith fell in love with Chamonix and was overawed by his first sight of Mont Blanc's splendour. Desperate to make an attempt on the summit Smith offered his services as porter to anyone planning an ascent, but to his great disappointment he had no one take up his offer – no small surprise as he was hardly experienced. He was even more disappointed to learn that Henriette d'Angeville made her ascent the day after he left Chamonix to return to medical school in Paris.

But so taken was he with this visit that he eventually gave up his medical studies in favour of writing and theatrical production. In 1848 he wrote and published a serialised book, *The Struggles and Adventures of Christopher Tadpole*, about a rather hapless man who climbed Mont Blanc, and in 1849 Smith travelled to Egypt and then produced a stage show about his travels called *The Overland Trail*, which apparently played to great acclaim in London. On the back of the profits from this show he returned to Chamonix in 1851 and actually climbed Mont Blanc, albeit in slightly grander style than Auldjo – Smith being the showman that he was.

Not one to travel light, Smith took with him on the ascent a large retinue of porters and guides who had to carry his substantial provisions, which included: 60 bottles of *vin ordinaire*, six bottles of Bordeaux, ten bottles of St George, 15 bottles of St Jean, three bottles of cognac, two bottles of champagne, four parcels of prunes, raspberry syrup, ten small cheeses, four candles, six packets of sugar, four legs of mutton, six pieces of veal and 46 fowl. Despite this lot, Smith managed the ascent and upon reaching the summit was said to have declared, 'The ardent wish of years was gratified, but I was so completely exhausted that without looking around me I fell down in the snow and was asleep in an instant.'[13] I'm not surprised as he was either completely exhausted or completely pissed. For any fans of *The Two Ronnies* out there, it goes without saying that the most satisfying item on Smith's list is 'four candles'.

Smith's ascent of Mont Blanc was both grandiose and somewhat vulgar, brash and showy in a way that he had made his own and it seemed that not only did he want to conquer the mountain but also conquer the myth of the mountain, expose what he saw as the exaggerated claims about how difficult and dangerous it truly was. In a sense, Smith wanted to stamp his unique mark on the mountain and turn his ascent in to another of his grand theatrical spectacles and upstage all who had preceded him. It is easy to understand how Smith was both admired and disliked; charismatic perhaps but the man was certainly wrapped up in his own rather narcissistic sense of personal grandeur.

As it was, the climb was not a walk in the park and Mont Blanc may have had the last laugh; its sharp aiguilles pricking the inflated balloon of Smith's overblown pomposity. Their group left Chamonix at 7.30 a.m. and at first all was fairly easy going as they reached the first stage of the Grands Mulets at around 4.00 p.m. Smith was confident and relaxed to the extent that he did not even bother with sleeping in his tent for that first night, opting instead to sleep out in the open under the Alpine sky as he watched the sun set and stars open up above him. As Smith wrote, '... the sight was more than the realisation of the most gorgeous visions that opium or hasheesh could evoke' and as the sun set the peaks looked like 'islands rising from a filmy ocean... an archipelago of gold'.[19]

But from here things started to get tougher for Smith and his party. The cold began to set in and Smith began to feel sick and languid and he was further disturbed by the sight of a young Irishman, who they had met on the climb the day before, lying in the snow vomiting and bleeding from the nose. Sending the man back down to the Grands Mulets, the group continued, with Smith dramatically recounting that every step they took was a step 'gained from the chance of a horrible death'. The temperature dropped further and Smith struggled to continue, having to be dragged onward by his trusty guides and an hour or so later on the steep slope of the Mur de la Côte he was staggering around, barely able to walk and demanding to be left alone to sleep. Much like our earlier friend, the Indefatigable Bourrit from all of 68 years earlier, Smith's latter account of his climb and the section on the Mur de la Côte was often prone to exaggeration and the over dramatic.

> *It is an all but perpendicular iceberg, there is nothing below but a chasm in the ice, more frightful than anything yet passed. Should the foot slip or the baton give way there is no chance for life... you would glide like lightening from one frozen crag to another and finally be dashed to pieces hundreds and hundreds of feet below, in the horrible depths of the glacier.*[19]

One of the group, a man called Floyd, said of Smith, 'He looked very ill indeed and was quite insensible when I poured a glass of champagne down his throat'. Whatever their pretensions, they certainly do things in style these Victorians; I hope Rupert will revive me in similar fashion should I fall by the wayside. I am sure he must have a bottle of bubbly stashed away in his voluminous rucksack.

Fortified and restored by the champagne Smith soldiered on, his account of the final stages of the ascent becoming increasingly lurid, no doubt fuelled by the unwise combination of happy bubbles and altitude sickness. Funnily enough, some people say that drinking a bottle of champagne is a good indication of what mild, low level altitude sickness feels like. Unfunnily enough, if you want to know what extreme altitude sickness is like then look no further than the recent experience of my work colleague who, on a charity climb of Mount Kilimanjaro, was left bleeding from her eyes, nose and ears and needed an immediate blood transfusion there and then. Amazingly, she still got to the top but can't remember a thing about it. And so, intoxicated by height and alcohol, Smith waxes lyrically:

> *... placed fourteen thousand feet above the level of the sea, terminating in an icy abyss so deep that the bottom is lost in obscurity; exposed in a highly rarefied atmosphere, to a wind cold and violent beyond all conception; assailed, with muscular powers already taxed far beyond their strength, and nerves shaken by constantly increasing excitement and want of rest – with bloodshot eyes and a raging thirst, and a pulse leaping rather than beating – with all this, it may be imagined that the frightful Mur de la Côte calls for more than ordinary determination to mount it.* [19]

At 9.00 a.m. on the 13 August 1851 Albert Smith, on his hands and knees, finally crawled to the summit of Mont Blanc, after being hauled up the final section by his guides. As said, on finally reaching the summit, Smith promptly fell asleep but awoke in time to knock back a few more glasses of champagne before making a hurried descent back down to the Grands Mulets, where they polished off the last of their provisions and then headed back down to Chamonix, returning to a hero's welcome that no doubt vanquished all memories of the arduous ascent. But that said Smith did not quite conquer Mont Blanc as he had set out to do. Indeed, the mountain had literally reduced Smith to his knees and perhaps, even for a small moment, knocked him off his self-appointed pedestal and cut him down to size. Smith would never admit to such a thing and in his later written account of the ascent gave an overblown and expectedly exaggerated description, airbrushing out the role of his guides and other members of the party, placing himself centre stage and leaving out the small fact that he was actually the last of his party to reach the top. Still, one would expect no less from the man.

When he returned to London, buoyed by the success of his earlier stage show, Smith put on his next theatrical spectacular, *The Ascent of Mont Blanc*, which first played to audiences in 1853 at the Egyptian Hall in London's Piccadilly. The show was a resounding success and ran for an amazing seven sell-out years and took close to £30,000 – no loose change back then. The show included an impressive diorama of the Alps rolled out across the back of the stage (not a million miles from the little moving panoramas Smith made as a child), as well as a couple of live chamois, a St Bernard dog, girls dressed in Alpine costumes and a large cut-out of a Swiss chalet. Over the top of all this rather chaotic sounding menagerie, the raconteur Smith would expound in his booming voice the exaggerated account of his ascent of Mont Blanc to the enraptured audience. The show, for all its brash tackiness, loose truths and blatant showmanship, gave the people of London a glimpse into an unknown world of cold, icy glaciers, dark crevices and the dramatic glory of summiting Western Europe's highest mountain and it must indeed have been a thrilling experience. In fact, so successful was it that Smith gave royal command performances at both Windsor Castle and Osborne House and when one of his St Bernard stars had puppies he presented one as a gift to Queen Victoria and another to a certain Mr Charles Dickens, who had also been to see his Mont Blanc extravaganza. Dickens said of the show, 'By his own ability and good humour, Smith is able to thaw Mont Blanc's eternal ice and snow, so that the most timid of ladies can ascend it twice a day… without the smallest danger of fatigue.'[1]

This then was the pinnacle of the Victorian fascination with the Alps and 1855 was the year in which the British people were, as *The Times* put it, 'gripped with Mont Blanc mania'. The road to the Alps had been well and truly opened, which meant more travellers, more tourists and more climbers. This was not to everyone's liking, with the likes of John Ruskin writing to the papers to complain that there had been a 'cockney ascent' of Mont Blanc, but this didn't bother Smith who realised he had tapped into a deep fascination of the time and was prepared to get all he could from it, which was a great deal. In the midst of all this mania, there was even a board game called The New Game of the Ascent of Mont Blanc, in which players, adults and children alike, could throw dice and progress along a series of 54 vignettes depicting Smith's adventure, starting from London's south-eastern railway terminus – a kind of Mont Blanc Monopoly. If luck would have you landing on the Mur de la Côte, your fellow players would each have to give you two counters in recognition of the extraordinary bravery shown in climbing it. I wonder if there was a square for having 'a glass of champagne poured down your throat' or one for 'being dragged on your knees to the summit by your guides'. Somehow I think not.

Thirteen

Today we have a 12-kilometre stretch and a height gain of a mere 860 metres, and by all accounts it should not be a too demanding stage. Also Kev Reynolds, our pocket guide, tells us that this section is one of the highlights of the TMB so we are in good spirits and looking forward to the day ahead. We have had a good rest day, really enjoyed our short stay in Courmayeur and feel ready to move on to pastures new, literally in this case. In fact there are a couple of route variants that we could take, one along the Mont de la Sax and the other along the Val Sapin. Each of these, although tempting, involve much greater climbs of around 1,500 metres, but the main route option looks like a great walk and we are happy to have a more relaxed day.

Breakfast done and rucksacks packed our little troupe of five departs the relative comfort of the Pensione Venezia and makes its way out of town. The identity of our group is now firmly cemented by the little furry marmots that Rupert bought for us yesterday and we have carefully secured them to our rucksacks. Except for Richard that is, who hasn't got one – a fact that he keeps reminding us of. Somehow though, even in his marmot-less state, he manages to declare himself official 'marmot leader'. He rationalises this with some nonsense about how the leader would not actually have to have one himself – the marmot denoting the sign of a follower. It feels like we are a real gang now and so taken are Claire and Julie with their fluffy friends that I would not be surprised if they insist that we do some kind of group marmot initiation ritual, perhaps a dance movement interpretation of the moment of awaking from hibernation.

The route leaves from the raised square in the centre of town and takes us up a narrow street between a church and the guide's museum that I visited yesterday. The street soon turns into the Strada del Villair and leads us out of Courmayeur and past tidy, well-manicured gardens and up into the little sub-village of Villair itself. Initially the surfaced road is no more than a gentle slope and provides a pleasant morning stroll but before long, and inevitably, the TMB route veers away from the road and takes us up through a steep section of woodland. This proves to be a long, hard uphill slog for a good couple of hours and in the rising heat of the morning we have to stop now and again to catch our breath. Once again it is turning out to be hot and sunny and we can't believe how fortunate we have been with the weather over the last five days. It is hard to believe that it could be anything else but perfect, so spoiled have we become.

The path is well established and easy enough to follow, but it twists and turns through the pine trees in a series of giddy switchbacks that reduces us to a slow and steady pace as we seek to conserve our energy. Eventually the trail opens up beyond the treeline into a grey-green field of grass and scree and leads us up to a small cluster of buildings and the Rifugio Bertone. The refuge seems closed but we stop for a rest and looking down the 700 metres or so from where we have just climbed, we can see Courmayeur spread out below like a little Lego town nestling in the basin of the valley, the mountains rising up on either side. At a spot just below the refuge there is a stunning view of Mont Blanc and the Aiguille Noire de Peuterey, the best view we have had so far, and it is simply breathtaking. The massif is a line of craggy peaks, standing proudly to attention, and above Mont Blanc dominates with its regal presence and crown of snow that dazzles in the bright morning sun. Its presence feels tangible, as if you could reach out a hand and touch the jagged, peaked wall of the massif. Glaciers stand out, frozen swathes of ancient ice and rock – retreating but still defying the heat of the September sun. A helicopter buzzes up from the valley below and hovers for a moment at eye level, like a reinvention of some kind of primordial, bug-eyed insect.

The refuge is perfectly located and Rupert and I agree that if we were ever to do this trip again, we might be tempted to skip the rest day at Courmayeur and hang out up here. The refuge itself was apparently built in 1982 to commemorate the mountaineer George Bertone, who died in a plane crash below the summit of Mont Blanc du Tacul in 1977. Another victim of the unforgiving mountains and a constant reminder of the dangers and risks that people take in these places. Reluctant to move on straight away we loiter for a while and load up with water, apricots and biscuits, topping up our energy levels after the taxing two-hour climb from Courmayeur.

Rested, we set off again and Claire and Julie are continuing with their project to seemingly document every bit of flora that they come across along the trail, stopping to snap away with the macro function of their very smart looking cameras. I am impressed with their diligence and knowledge and the simple excitment and joy they express with each flower or plant they discover. Carried along by their enthusiasm, I dig out my own camera and discover that I too have a macro function and soon find myself also stopping every few minutes to take a picture of a grasshopper. I have borrowed my camera from Nicky and this is my one concession to technology on this trip. When on the GR10 with my friend Rob, we made do with disposable cameras that we posted home along the way in an effort to keep our weight down. But this time around I decided to temporarily go digital – but that is as far as it goes.

That's the thing with middle age and technology; you reach a kind of tipping point where you have to make a choice about the extent to which you embrace or reject technology. Perhaps reject is the wrong word – more allowing it pass you by, like a digital cloud on a hazy day. Back home I glance around at my friends and colleagues who are mostly all hooked up to their iPhones, iPads, iPods and iWhatevers. Book groups now consist of people sitting around clutching their Kindles (although if you are reading this on a Kindle you are forgiven) and every few months it seems a new smartphone is unleashed upon the market, not to mention the video game companies who very insidiously manipulate the market to leave young children clamouring for the next big thing.

Everyone seems very excited by technology, but what exactly is the big deal I wonder? Of course it is all about speed;

faster, faster, faster and we all know that faster means better. Doesn't it? I realise that as I approach 50 I have to accept that I am out of the techno loop and am in danger of coming across as a curmudgeonly Luddite, viewing life through my nostalgia-rimmed, rose-tinted specs. As Douglas Adams once said, 'Anything invented after you're thirty-five is against the natural order of things.' Well, I guess that's about right, but let's just wait and see, because I have a sneaking suspicion that this exponential growth in our technological capability and the collective hunger and desire to consume this technology is going to drive us up a one-way street so fast that we will never quite know how to get back to where we were at the beginning. In other words, we will lose our way and, to be dramatic about it, lose all that is good about the human condition. The cure? A long walk in the mountains I would say.

Many say that the Internet is a force for good; a wealth of knowledge at our fingertips that can only enhance and enrich our lives. Well, close to a third of all Internet traffic is dedicated to pornography and it seems to me that much of the rest of it involves wobbly phone footage of people running around in parks shouting 'Fenton, Fenton' or sticking their heads into ceiling fans for a laugh. Is this progress? We have a 9.00 p.m. watershed for TV programming, while at the same time children as young as seven and eight play on computers in their rooms just two or three clicks away from all the hard-core, violent pornography that any of us would ever want to imagine.

In the context of my work with very traumatised children I have long been calling for the Internet to be regulated and it is interesting to see that only now are people beginning to wise up. All I am saying is that we need to be a little more discerning in our understanding of the impact that elements of these new technologies have on our day to day lives, and especially the lives of our children. We can't afford to engage in a grand social experiment with the next few generations before realising it was all a terrible mistake.

As you might have gathered I feel strongly about this, so please excuse the rant, but that's because I see its impact every day upon the children I work with. And what on earth does all this have to do with walking around Mont Blanc, you may well ask? Well, quite a lot really. One of the most pleasurable things about long distance walking is the absence of everything online. No Internet, no email, not having to be accessible or contactable at all times of the day. No screen time, which is bliss. The walk allows the time and space to re-engage and reconnect with the world, to slow down and tune into the natural rhythms of life, which for so many of us are obscured by the unnatural intrusions of modern life. I would gratefully swap the washed out perma-glow of my computer for the orange glow of the morning's first light on the grey-blue peaks; the rippling contours of the Alpine landscape that shifts with the shadows as the sun moves across the sky; the fresh smell of pine and the gentle sound of cow bells, ringing in the end of another day; the simple physical sensation of one's foot upon the stony ground and the joy of physical exertion. These are the sensations that invigorate the spirit, stimulate the mind and ultimately make one feel alive.

I admit, I am something of a technophobe. We drive a Morris Minor and a VW camper van. We have an antique 1920s phone that fills the house with the sound of a real hammer on a real bell. We have a wind up gramophone that we bring out at Christmas. We even have our milk delivered to the door in real milk bottles. Do these things improve my quality of life? Yes, absolutely. I don't own a smartphone and never will; I have a 20-year-old vintage Nokia that serves me fine.

Perhaps I am being over cynical when it comes to the wonders of modern life and must remind myself not to be to negative. I do tend to have a rather poor world view. I am the one with the half-empty mug after all! I recall, once again on the same beach in Cornwall one New Year's Eve, when Rupert brought out some lanterns for us to send up in the sky, one for each of us that would carry away into the New Year's heavens our hopes, dreams and ambitions for the forthcoming year. We watched in awe as Nicky's, Jess' and Rupert's lanterns floated glowingly and beautifully into the night sky. My lantern took a few shuddering leaps into the cold air and then crashed and burned into the freezing water of the Helford estuary like a mini Hindenburg. As we watched it head inexorably for the water I remember turning to Rupert and saying 'That's my life, that is.' I have never seen him move so fast as he ran to avert the inevitable lantern disaster, knowing that my very future lay in his hands and that he was responsible for this terrible omen. I remind him of it now and again when he is looking too chirpy and indeed one might have thought he had learned his lessons about flames and omens on Cornish beaches at New Year. Enter stage right: Toby the smoking dog.

Anyway, despite my aversion to technology I am enjoying the macro on Nicky's camera while Rupert has discovered to his anguish that he is macro-less. Richard is not too bothered with any of this and is more intent on working the TMB grapevine for any useful snippets of information that he can assimilate into his hive mind. Perhaps that's it. Richard is the Borg of *Star Trek*. Perhaps we will all be slowly assimilated into his ever-expanding TMB collective – whatever that means. Earlier, Richard actually referred to himself as a TMB tart and then said something about pimping that I didn't quite catch. I wonder if he is on an entirely different trip to the rest of us. Since being in Italy, Rupert and I have renamed Richard 'Ricardo', which seems a bit more gangster-like and much more apt for a man of his stature.

The path from the Rifugio Bertone takes us on a sweeping contour around the broad edge of Mont de la Saxe and opens up to some incredible views towards Col de la Seigne, Mont Blanc and the Grandes Jurasses, a great sweeping wall of rock that supports a dramatic series of glaciers, great dirty wedges of ice that look as if they have been painted onto the face of the rock with a rough stroke of the brush. With the Val Ferret falling to our left the trail levels out as we traverse the flank of the Mont de la Saxe, which rises up to our right. The walking is easy going, but the temperature really picks up as we hit the middle of the day and it feels more like mid-August than mid-September. Buoyed by the sun and the ease of the walking, our little group feels cheerful and relaxed. The girls are chipper and their natural interest and curiosity in the world around them is pleasantly infectious. Ricardo is dour and self-deprecating in an amusing way, and he continues to regale us with stories of his many global adventures. Now and again someone calls out for a 'marmot stop' where we have to check that our marmots are still securely attached to rucksacks, all present and correct.

As our guidebook said, this is indeed one of the high points of the TMB so far. With the great sweep of the Val Ferret dropping dramatically away to our left the pathway serves as a spectacular balcony, looking out across to the great barrier of rock that is the Mont Blanc massif, the scale of which is almost beyond words as it rises up almost vertically from the valley floor and dwarfs us as it continues ever skywards. In the clear warm air, the illusory sensation of proximity is tangible; it feels close enough to hold a hand out to, yet far enough away to be hardly able to pick out the tiny dotted presence of climbers on the far mountainside, barely visible to the naked eye. The peaks of the Dent de Géant, Dôme de Rochfort, the Grandes Jurasses and, of course, Mont Blanc itself all tower over us, all part of this great range that pushes out of the ground like the gnarled backbone of some vast slumbering beast. Travelling northwards from this point over the mountains, as the crow flies, one would come more or less to Les Houches, our starting point on the TMB and it is strange to think that we have reached close to our halfway point, that we have walked this far in what feels like a comparatively short amount of time.

The trail itself is a rich, intoxicating cocktail of colour and smell and sound. The vibrant green pastures are dotted with pine, fir and larch – the fresh scent of the trees carried in the warm breeze like alpine elixir – and the trail occasionally takes us through wooded areas of birch, maple and oak. Here and there mountain ash trees – the rowanberry – add impressionistic dapples of vivid red as their autumn fruit hangs in abundance, a gift for the local wildlife. Juniper and bilberry texture the ground and despite the time of year, a variety of flowers mark our route with little splashes of colour, much to

Claire's and Julie's delight. The contrast between the rich green of the meadows and the rippling grey of the mountains beyond is striking and looks almost unreal at times – a piece of Albert Smith's panoramic scenery rolled out onto the stage like some kind of vast theatrical backdrop.

Sometimes our group walks together in silence, each of us wrapped in our own thoughts, and sometimes in occasional chatter and despite the fact that we have only known each other for a few days there is an easy companionship. We are joined in our endeavour, the shared aim of walking around this mountain and in many ways that is enough – we don't need to know much more about each other. Our backstories are of little consequence; it is within the present that our quest has meaning as we temporarily join in this common experience.

We come across a wonderful spot and decide to take a rest for lunch. A spur of meadow pushes out, forming almost an overhang above the valley, and the grass weaves a rich carpet on which to sit and reflect for a moment. Mont Blanc is perhaps as close to us here as it ever will be, certainly it feels that way, as we sit in this little north-western nook of Italy that rubs frost-bitten noses with neighbouring France and Switzerland, both of which we look into from this vantage point high above Val Ferret. It is as an idyllic and peaceful spot as one could wish for as we share a lunch of ciabatta, cheese and some spicy sausage cadged from Ricardo. We hang around idly in this place for an hour or so and Rupert and I chat about work, therapy, fathers and, for some reason, piles. I can't quite recall why we talked about piles, but it is something to do with the bidet that was in our room back in the pensione in Courmayeur that I threatened to use.

There is something special about this spot that makes me feel reluctant to move on, although I am unsure why. Rupert and the others pack their rucksacks and prepare to continue walking and I tell them to go ahead and that I will catch them up later, using the pretext of a having to make a phone call home. I sit for a while on my own, simply soaking in the intense tranquillity of the moment, the stillness of the air, the mountains towering over me, the glaciers that hang supported in the rock face, dirty white rivers of ice and the warmth of the September sun on my back.

I feel at peace here, unusual in the context of my usual restlessness, and I feel a little sad as well. Alone, upon this spur of rock and grass that hangs in the air high above the valley below and low beneath the mountains above, I fish around in the top compartment of my rucksack and pull out the small plastic vial containing some of my father's ashes. I slowly unscrew the lid and leaning outwards towards the mountain wall on the far side of the valley, give the container a flick of the hand and send the ashes into the air, tiny particles of dust, powdered memories. In the stillness, a brief gust of wind appears from nowhere and carries the ashes away.

Perhaps it is the stark contrast of height and depth that makes me think of my father, the mania and depression that was so much a part of his life and mine. The impenetrable rock, the inscrutable paradox of a landscape that can be at once both harsh and inviting. The beauty of the place that I know he would have relished; the proximity to France and Switzerland – places that he knew so well. I am not a spiritual person, far from it as I have said, but I recall that on the same morning of my father's death I had a powerful and vivid dream that he had in fact died. When I woke up I felt so troubled and disturbed by the dream that I told Nicky about it. 'Why don't you give him a call?' she said and so I did, but there was no answer. I called him several times that morning and there was no answer. A couple of days later my mother discovered his body in his flat, on the floor where he had fallen after a massive heart attack. Now and again I think about the phone ringing in the empty silence of his flat, his body lying on the floor. If a phone rings and there is no one there to hear it, does it make a sound? I pick myself up off the soft grass, stuff the empty tube back into my rucksack and with a brief nod in the direction of Mont Blanc prepare to head off to catch up with the others. As I haul my rucksack onto my back I notice my little marmot lying in the grass, having somehow become detached from the clip. I carefully reattached the little creature and wonder what, if anything, its near loss might have symbolised.

Fourteen

Wandering Jew that he was, my father was always on the move – never quite content – and as I have said, it's a trait that I recognise in myself; the familiar nagging of 'the grass is greener' syndrome. I am always banging on at home to Nicky and Jess about the need to move away to the countryside and escape the clutches of suburbia and indeed we have got close several times, only to fall at the last hurdle for all manner of reasons. That's the thing about suburbia, once immersed in its initially inviting des-res depths it is hard to haul yourself out again – like a warm bath on a cold winter's evening. There is something quite insidious about the benign suburban veneer of comfort and ease that induces a middle class torpor, an existential fug that can last for years and from which it is really quite hard to escape. So when I am walking in places like this I feel inspired into action, but then once home soon slip back into its comfortable clutches. The thing about suburbia is that it is nondescript, you could be anyone and anywhere and over time it slowly but surely eats away at your soul, sapping your will to live and gradually turning your brain into a pebble-dashed, privet-shaped mush.

Walking in the mountains makes me feel alive, a temporary escape, and for many years my fantasy has been about one day buying a little place where I can hide away and live the life of a hermit. As my father used to say, one should never live where the open countryside is more than two minutes from your front door, with which I couldn't agree more. Another of his wise sayings was that his idea of hell, being a man of books, was a bedroom that had only one light switch in the corner by the door. True as well. He also said that all social workers should be shot, just after I proudly told him that I had qualified as a social worker after two years of hard labour. That perhaps was not one of the high points in our relationship, bless him. Anyway, the funny thing is that instead of moving myself I seem to have spent an inordinate amount of time moving other people, perhaps in the thought that if I do it enough I might one day gain enough momentum to take the plunge myself. Or perhaps I just vicariously live out these fantasies through others. Who knows?

Back on the trail, we continue on towards our destination for the day, the Rifugio Bonatti. The path is well worn and undemanding, stony in places but an easy track to follow. Perhaps because of the option of the other route variants, we have seen less people today and the fact that we have been dawdling and taking our time means that most people are probably ahead of us. For long periods we don't see anyone else and it is refreshing after the crush back at Elisabetta and the relatively busy Courmayeur. At one point a group of five British guys come storming up behind us and as they bustle past they tell us that they are attempting the whole of the TMB in just six days, doing two stages rather than one on some days. Certainly it's possible, but why? It seems nonsensical; why rush such a wondrous experience? For us this trip has been all about slowing down and it seems strange that these folks should want to speed up to the point where they hardly notice what is around them. To emphasise this, as they overtake and stride on forcefully ahead one of them misses his footing and goes flying head-first into the bushes by the side of the track. It is clear that they are tired and pushing themselves too hard and it is easy to lose concentration and miss a step in that state. Fortunately the chap was fine but on another section of the path, much of which has steep drops to the side, his fall could have been very serious. It's questionable to push yourself like this at the cost of personal safety, but we wish them well as they stride ahead into the distance.

Occasionally we pass ruined farm buildings, timeless in their derelict state with weathered, timbered beams pushing out through tumbled blocks of pale green, lichen-covered stone. Some of them would certainly provide temporary refuge if required and I would guess that these places might be frequented by local shepherds now and again when the weather suddenly turns. The path leads down into the Vallon d'Armina, through which flows a small stream from the Col Sapin that sits at 2,436 metres high up to our right. Physically, I have been feeling pretty good. Although the regulation morning ascents that generally form the pattern of each day's walking have been tough and arduous, the physical and mental exertion of the climb soon evaporates, and today's section has been a delight, much of it a gently undulating and undemanding traverse along the flank of the Mont de la Saxe. My rucksack feels fine and measurably lightened after dumping a pair of thick shorts and a couple of books back at the Rifugio Elisabetta. Rupert, however, is still carrying half his house on his back although he too lightened his load by taking the opportunity back in Courmayeur to post home some of his more weighty items, namely his watercolour paints and some unneeded clothes. All the same, we reckon that Rupert's rucksack is almost twice as heavy as the rest of ours and I think he is feeling it at times.

His blister seems to have calmed down now and he has finally settled into his walking boots so all seems well on that front. My own boots have been fine. These are my trusty Berghaus boots that saw me through 850 kilometres on the GR10 without a single blister and although the stitching is starting to go in some of the seams, I can't imagine walking in anything else. I have continued to keep up my regime of applying Vaseline to my feet every morning, a tip passed on by Rob, my GR10 walking companion, and I swear it is one of the best bits of walking advice I have ever been given.

After about seven hours on the trail, allowing for our rather leisurely pace and lengthy breaks along the way, we catch the welcome sight of the Rifugio Bonatti that we approach up a stiff grass slope on our right. The refuge is a new, solid looking building constructed from stone and timber and sits proudly on a flat area of ground above grassy slopes and has a spectacular *en face* view that embraces almost the entire mountain range from Col Ferret to Col de la Seigne. Outside, the picnic tables provided by the refuge are taken up with numerous walkers, climbers, mountain bikers and several of our TMB associates, who now greet us like long lost friends. The two South African ladies we met back at Elisabetta (practically slept with I should say) are delightful and form a friendly reception committee as they clap and cheer our arrival. It is a relaxed, laid-back scene as everyone soaks up the late afternoon sun and there is an immediate feeling of camaraderie through shared endeavour that ripples warmly around the group.

Inside, the refuge is a perfect example of how these places should be run. Spacious and well organised it has the capacity to absorb the constant stream of hikers that flows down the valley and even though it sleeps close to 80 people, and must be close to capacity today, it somehow does not feel overly crowded. The place is well furnished and comfortable and along with the regulation larger dormitories it also has smaller two- and six-bedded rooms for those shrewd enough to book ahead. We are shown up to one of the large dormitories and even though the beds are still laid side-by-side in great lines there are no bunks and the extra space makes it almost seem civilised. The water in the showers is plentiful although in my experience cold, but this could have been because I did not quite get the hang of their token system.

After showering and ablutions and the now de rigueur rummaging in rucksacks for no apparent reason – searching for things that we know are not in there in the vain hope that we might, with a theatrical flourish of the hand, produce a gin and tonic and a copy of today's *Guardian* – we head back outside and sit at the tables to chat, write, drink beer and await the much anticipated sunset. And we are not let down. As the sun begins to fall behind Mont Blanc, that sits high to our northwest, the scene changes magically as transient shafts of orange and red sunlight subtly paint the few available clouds with dappled brush strokes before beginning to slowly spread across the ever darkening sky. The mountains seem alive; the contours, folds and veins of rock, water and ice shifting and rippling in the changing light and the painterly qualities of texture, form and depth suddenly reveal themselves, a seductive glimpse of a softer, more gentle side of this range. Among the scattered groups outside the refuge there is a hushed sense of reverence, a collective feeling that this is a special moment and one not to be interrupted with casual words. Many of us take photographs, all in the unsaid knowledge that no photo can do this scene justice, and that it is an experience to be 'felt' as much as it is to be witnessed.

Later we gather in the large restaurant-like dining area where we are randomly thrown together around long wooden tables upon which bowls of hot food are casually distributed and hungrily devoured. We eat egg fritters of some kind with plenty of carrots and potatoes and it is good to have some fresh vegetables. In fact, the food in these places is invariably good and based upon large amounts of protein and carbohydrate – fuel for the long distance hiker. We are sat with a group of Americans from Washington State and most of the talk is about their encounters with bears during various camping and hiking trips in the US. Both Rupert and

I are constantly taken aback not just by the numbers on the TMB, but also by the vast international spread of our fellow walkers. Indeed we Brits are in the minority most of the time. Claire and Julie, sticking to their budget, are cooking outside with their trusty camping stove and later we rejoin them to take in the last of the dying light, the mountains now like great towering shadowed creatures; a benign but powerful presence in the gathering darkness.

Ten o'clock seems to be curfew time, and feeling physically and emotionally sated from both the marvellous meal and the stunning sunset we retire for the night, crawling into our beds in little rows of humanity within this mountain ark of a refuge, as if we are about to embark upon some great sleep-induced interplanetary adventure to colonise a new galaxy. Or perhaps I have just drunk too much wine. Perching my marmot carefully on the wooden beam above my head I dig out my notebook and torch and attempt to write up my journal. Rupert and I whisper conspiratorially in the darkness and for some reason he asks me to tell him a story, so I make up something about a marmot that has lost his nuts. Fair to say, it's not great as stories go and without any significant metaphorical conclusion or indeed any inventive plot twists it simply remains a rather basic narrative account of a marmot looking for his nuts, which perhaps is metaphor enough for some people, thank you very much. In the darkness of the room someone makes a 'shh' sound and I realise I must have been telling the story to the whole dormitory. Had I known they were all listening in I would have made more of an effort, especially as my reputation on the TMB is still suffering from the great Elisabetta debacle. I whisper Rupert a poem I wrote earlier, but he is not too impressed and seems much more interested in why I have carefully perched my marmot up on the wooden beam above my head like some kind of hairy, rodent guardian. Perhaps he is wondering where its nuts are.

Fifteen

The Rifugio Bonatti is named after Walter Bonatti, the Italian climber, explorer and writer, and striking black and white pictures of this charismatic, rugged looking man adorn the walls of the refuge, a tribute to his great accomplishments. Bonatti died only recently in fact, in 2011 at the age of 81 and was heralded by many as one of the finest alpinists and greatest mountaineers of all time, although his life and career were not without controversy. 'I have often wondered,' Bonatti wrote in his memoir *The Mountains of My Life*, 'whether I was born a loner or became one.'[14] Perhaps the capacity to function alone is an integral element of mountaineering psychology, and the process of climbing is an activity that requires long periods of intense mental concentration and physical challenge in regions of dramatic isolation where each decision is of vital consequence. It also requires supreme trust in the capabilities of others, which Bonatti discovered to his cost. He undertook many of his greatest and most challenging climbs alone, maybe because that is how he preferred to operate. But it may also be that he climbed alone as a response to a climbing community within which he had for many years felt maligned, outcast and poorly treated.

The controversy came in a 1954 first attempt on K2, the second highest mountain on the planet after Everest. Bonatti was the rising star of the mountaineering world and had clearly earned himself a place (although perhaps not welcomed) on the Italian team alongside Lino Lacedelli and Achille Compagnoni and a Pakistani porter called Amir Mahdi. As the expedition reached its heady climax, Bonatti and Mahdi were required to take critical oxygen supplies up to camp IX, the final camp from which Lacedelli and Compagnoni were poised to make their final push for the summit. However, all did not go to plan and the shelter at camp IX was not where they expected it to be and as the darkness gathered around them Bonatti realised that the other two were still some distance above them, Compagnoni having decided to set up camp at a much higher location than had been previously agreed. Lacedelli and Compagnoni shouted down for Bonatti to leave the oxygen and go back down, but the porter Mahdi's physical condition was deteriorating rapidly and he was in no shape to either continue on or make it back down to camp VIII, so the pair had no choice but to endure the night in an open bivouac without tent or sleeping bags at an altitude of 8,100 metres and minus 50 degrees Celsius. Mahdi lost his fingers and toes and Bonatti was lucky to survive the night unharmed.

Lacedelli and Compagnoni went on to reach the summit, using the oxygen that Bonatti and Mahdi had brought up, and Italy celebrated this epic achievement while Bonatti's contribution to the expedition was overlooked and he found himself ostracised by the rest of the team. Bonatti accused

Lacedelli and Compagnoni of purposely changing the location of the final camp IX so that he would not be able to reach the summit with them, endangering the life of both Mahdi and himself in the process, perhaps even of trying to kill them. Years of bitter recrimination and in-fighting ensued, with Lacedelli and Compagnoni counter-claiming that Bonatti had used up some of the critical oxygen, somewhat ironically as Bonatti was the youngest and fittest of the team and the most likely to be able to summit without the need for oxygen, a feat that would certainly have eclipsed that of Lacedelli and Compagnoni, which was perhaps what they feared. Later, Bonatti planned a second attempt on K2, but was frustrated and blocked by the ongoing fallout of the 1954 expedition.

Bonatti was vilified for years to come by many in the climbing community and it was not until 53 years later that it was finally accepted by the Italian Alpine Club that Bonatti's version of events was accurate, although it was a controversy that bitterly haunted Bonatti for much of his life. Reinhold Messner, perhaps the greatest mountaineer in history, said in June 2011 in an article in *La Gazzetta dello Sport* that Bonatti 'leaves a great spiritual testament: he was a clean man vilified for 50 years over what happened on K2, but in the end everyone accepted that he was right'.

Many years later, Bonatti wrote in *The Mountains of My Life*:

> *Until the conquest of K2 I had always felt a great affinity for and trust of other men, but after what happened in 1954 I came to mistrust people. I tended to rely only on myself. This was limiting*

me and I knew it, but at least served to protect me from further disappointment.[2]

Maybe he was born a loner, who knows, but certainly his experience on K2 set him apart from others and it is easy to understand his preference to climb alone from then on.

One of his many great solo climbs was in 1955 when he forged a new route on the southwest pillar of the Aiguille du Dru, the sharp mountain pinnacle that lies east of the village of Les Praz in the Chamonix valley. The ascent took Bonatti six days and required five hanging bivouacs, literally a platform shelter that hangs in the air from a secured position in the rock above. Even today this climb is considered a masterpiece of mountaineering endeavour. Apparently, after five days of climbing on the vertical rock face Bonatti became stuck at an impassable section of overhang that he couldn't negotiate; the rock on either side of him being absolutely smooth with no possibility of a handhold. Tying together all the slings and bits of rope that he had with him, Bonatti secured one end in a crack in the overhang above him and then swung back and forth from the other end until he managed to find a hold so that he could continue. He did all this from a height of around 3,500 metres! Subsequently, this route became known as the Bonatti Pillar and was a test piece for climbers for years to come until it was destroyed by a large rockslide in 2005.

Whatever the circumstances, a mountaineer's career is inevitably going to be marked by points of tragedy and Bonatti certainly had his fair share. One of these times was in 1961 when he and a team of two others attempted an ascent of the Central Pillar of Frêney, one of the hardest and at that time an unclimbed peak within the Mont Blanc massif. During the approach they met up with a French team of four and the two teams decided to embark on the ambitious ascent together. However, the fearsome elements conspired against them and a severe snowstorm lasting for more than a week halted their progress just 100 metres from the summit of the pillar. Unable to move for three days they finally decided to descend, but only three of the seven climbers, including Bonatti, managed to return safely. The remaining four died of either exhaustion or accident as they tried to find a way back down through the extreme conditions. Latterly, in 2002, Bonatti was awarded the National Order of the Legion of Honour by Jacques Chirac, then president of France, for the 'courage, determination and altruism' he demonstrated trying to save his fellow climbers. On the 29 August 1961, just days after Bonatti's attempt, the Central Pillar of Frêney and the 'last great problem of the Alps' was finally summited by a group that included Chris Bonnington, the renowned British mountaineer.

In 2009 Bonatti became the first climber to receive a Lifetime Achievement award by the mountaineering community, well deserved if belated recognition for his incredible career. And so the Rifugio Bonatti stands as another tribute to this great alpinist, the building looking out towards the dramatic southern flank of the Mont Blanc massif, his spiritual home and site of so many of his extraordinary feats of physical endurance.

At about 6.00 a.m. I am gently woken up by Rupert who tells me he is going out to watch the sunrise and asks whether I want to join him. The temptation to simply close my eyes and drift back into the beguiling dream-world of early morning dozing is strong, but I fight this off in the knowledge that this is one of those potentially special moments that only come our way once in a while. Outside, quite a number of other people have made the effort to get up early and we stand in contemplative, reverential silence in the cold, dawn twilight – the sun yet to make an appearance above the rocky horizon to our east. Over the course of the ensuing minutes, the quality of twilight subtly yet perceptibly changes in preparation for the sun's introduction and the feeling among the group also palpably changes in the knowledge and expectation that this truly is a special encounter with the natural world. Absorbed in the moment, we become attuned to the slightest change of light and then suddenly, with a collective ripple of acknowledgment within the group, the sunrise has begun.

With Mont Blanc being pretty much to our west, its domed summit is the first to be picked out by the red-orange glow of the morning's light, a faint hint of the sun's majestic power. Then other sections of the great southern flank of the massif are highlighted by this magical solar marker pen; the outcrops, ridges and contoured crags of the range that catch the light create a wonderful rippled perspective as the light shifts quickly from red-orange to orange-yellow. The deep floor of the Val Ferret below us remains entombed in shadow and it is as if two worlds exist – the shimmering golden castles in the air that float mystically above some kind of subterranean underworld; light and shade, conscious and unconscious. I am not religious, but there is something almost spiritual about this spectacle, certainly magical, and one can't help but reflect upon the thousands of years that the sun has risen over these mountains and the people who have witnessed this vision of nature in times gone by. What did they make of it I wonder?

Somewhat overawed by the experience, we return inside and join those who are already having breakfast and busily preparing to set off for the day. Compared to the serene nature of the sunrise outside it feels noisy and intrusively bustling inside and in an attempt to get myself back up to speed I drink three cups of strong coffee in quick succession, soaked up with some cereal and toast. Apart from the early morning start, I slept reasonably well during the night, only occasionally disturbed by Ricardo's snoring next to me. I can't bear to tell him though; to be labelled as a snorer is a crime on the TMB and we know what happened to Rupert the last time I said anything about snoring.

Today's section is a fairly lengthy 20-kilometre hike with a total height gain of around 900 metres and then a potentially long and arduous descent of 1,400 metres that takes us up and over the Grand Col Ferret, out of Italy and into Switzerland to our end point for the day, the small Swiss village of La Fouly. Our pocket guide, Mr Reynolds, has it down for about six to seven hours, but as we have come to realise these time estimates are somewhat arbitrary and we are not the fastest of walkers. There is also a lot of pre-departure commotion going on in the reception area of the refuge, which seems to be mostly about tonight's accommodation options. It would appear that the refuges in La Fouly are full and everyone is desperately ringing around trying to find a place to stay and then to factor this into their route options for the day. It is all a bit frantic and the anxiety is infectious as all sorts of places, options and possibilities are explored. A friendly English guy who speaks fluent French rings a couple of places for us and it seems that the only possible place to sleep tonight is an expensive hotel in La Fouly, way beyond our budget. Still, we tentatively book a room for Rupert, Ricardo and myself while Claire and Julie make alternative plans.

We leave the Rifugio Bonatti at around 8.30 a.m. and the going is initially very pleasant as we traverse the side of the valley along a well trodden path, crossing streams as we go and then begin to slowly drop down towards the floor of the valley. We pass more photogenic, derelict farm buildings and it is a joy to walk through the green pastures of the undulating hillside, scattered with pink clusters of alpenrose (a type of evergreen rhododendron shrub) and more of the bilberry, rowanberry and larch that we passed through yesterday. It's an evocative alpine scene, the sprinkling of pink, blue and red across the rich, verdant carpet of pasture and larch and all the time the ridged grey presence of the mountainside laced with vertical white ribbons of glacier. The snow of the Grandes Jurasses, 4,208 metres high to our left, glints in the morning sun and the deep azure of the sky completes the spectacular canvas. Once again we are blessed with a fine sunny day and as we walk we see a helicopter making low passes up and down the valley. We wonder if it might be a sightseeing trip, but then a little later we are greeted by a mountain rescue worker with a photograph of a man who has gone missing somewhere in the area. It is a sobering moment, the thought of this missing man, possibly injured or even dead while somewhere his anguished family and friends desperately count the hours as they wait for news. In the benign blue skies and warm sun of this lovely September day it is easy to forget how perilous these mountains can be; that people regularly get lost, hurt, killed; and that the weather can turn on a sixpence when you least expect it. We wish the rescue worker good luck and continue on our way, feeling a little sombre after this encounter.

A bit later I realise that I have lost my marmot. It must have somehow become detached from my rucksack and fallen on the ground somewhere between here and the Bonatti refuge. It's an odd coincidence that I nearly lost it back at the point where I scattered my father's ashes and now have actually lost it after last night's story of the marmot that lots its nuts. Perhaps the marmot was more symbolic than I thought. The girls are devastated and very touchingly suggest that we launch an official search party, retracing our steps back to the refuge, or that we could show a photo of it to passing walkers. But I feel OK about the loss, perhaps it was meant to be, and we continue on our way.

Before long the path drops down to the valley floor, where we emerge beside the Chalet Val Ferret, a bar/restaurant that also provides accommodation for up to 14 people. The chalet sits at the limit of the driveable section of the valley and so provides a good base for walkers and day trippers who want to explore the surrounding hills. From here we follow a tarmac path for roughly a hundred metres before crossing a river and then begin climbing again, steeply at times, the sudden incline something of a shock after the ease of the walking so far. After about 45 minutes the dirt track takes us up through the treeline and we arrive at the Rifugio Elena, where once again our South African lady friends form a welcoming reception committee and cheer and clap our arrival. These two amazing women must be in their mid-sixties, but they have bundles of energy and we can't quite work out how they seem to be ahead of us each day, settled in and awaiting our arrival as if they have all the time in the world. It is as if they have been beamed down from the Starship Enterprise. Their enthusiasm and simple delight in the moment is infectious and even a little humbling – there are times when I would welcome just a little of their *joie de vivre*.

The Rifugio Elena sits at 2,062 metres and is a solid, functional, sturdy looking place built into the hillside, which is probably for the best considering that the original building was wiped out by an avalanche in the 1950s and subsequently rebuilt in 1995. The refuge is apparently dedicated to Queen Helena of Italy who married Victor Emmanuel, King of Italy from 1900 to 1946. However, the people of the refuge like to tell another story about a young shepherd girl who once lived with her father in a small shelter in the local hills. The little shepherdess knew nothing of the richness of the royal court, but it was said that she owned the twinkling of the stars and the light of the moon that could transform ice into precious jewels. The days were long and hard and the poor girl was struck down with an illness that eventually led to her death. Suddenly, everything began to dim: the mountain peaks lost their clarity and the dew on the grass no longer reflected the morning light. But nothing in the valley had actually changed; it was the eyes of the girl's father, the shepherd, dimmed with the tears of his grief. He left the Alps, but not before he named their little shelter after the princess of his heart, his little girl Elena.

The place is abuzz with activity: people breaking their journey for a bit of light refreshment, school groups on a day out from the valley, day trippers having lunch. We stop for a while and have a drink and eat some of the packed lunch that we bought back at Bonatti. The refuge has a large terrace that looks out across the valley to the Glacier Pré de Bar, which runs down from Mont Dolent. It is certainly a good view and we can see the upper basin of the glacier, rimmed by the ridge of the surrounding peaks and the dirty grey tongue of the glacier itself, squeezed ingloriously between two banks of rock as it pushes down into a grey bowl of moraine, scree and mountain debris. In fact, when I say 'pushes down' it might be more apt to say 'once pushed down' as this glacier, like most of those in the Alps, is in retreat. A study in 2011 found that in the ten years from 2000–2010 Alpine glaciers lost on average more than a metre of thickness each year. The Mer de Glace, France's biggest glacier, has retreated by a kilometre in the last 130 years and thinned by 27 per cent. A study of Swiss glaciers in 2008 found that 78 were retreating, two were stationary and five were advancing.

There is no doubt that overall the glaciers are in retreat, in fact more than half of the ice-covered areas and probably two-thirds of the ice volume in the Alps has disappeared since 1850. There is a clear consensus that the glaciers are shrinking and that this rate of shrinkage is increasing, with some fearing that these great ice giants could disappear within the space of a generation. There is less consensus about the precise reasons for this rate of shrinkage. Most scientists say that glaciers are very direct indicators of climate change and while factors like precipitation and wind clearly play a part in this process the general view seems to be that rising temperatures are the main cause. As to why the rising temperatures, well that's another story, but the science on this is unequivocal and I know where my money lies.

Anecdotal perhaps, but the people who live and work around Mont Blanc are very clear about the fact that the glaciers are in recession, the mountains are warming up and the physical evidence is all too apparent. We can argue about the cause, and I am sure that people will, but environmental scientists predict that the impact of global warming on the Alps will certainly be felt by 2050 – if not sooner – and while some may feel this is still some considerable time off, it is now that politicians need to make decisions to confront the inevitable challenges ahead. There seems little doubt that climate change will have a considerable impact upon the main economic activity of the Mont Blanc region, and indeed throughout the Alps. Less snow on the lower altitude ski slopes will affect the tourist and leisure industries and it may well be that agriculture, natural habitats, river patterns and the regional flora and fauna could all be radically transformed as well as possible increases in natural hazards like floods, landslides and avalanches. I would suggest that this is no time to be sceptical or to stick one's head, ostrich-like, in the snow. Only time will tell.

Sixteen

The September sun continues to beat down and whether or not this is the result of climate change, we welcome it and actually have to seek out the shade to protect us from the heat. We fill up our water bottles and prepare for the big push up to the Grand Col Ferret, which judging by the path that we can see snaking its way up the steep hillside behind the refuge, is going to be quite a climb. The honour of 'king of the hill' has been shared between our group, depending upon whoever it might be that day who gets into the zone, the groove – their walking 'flow state' as one might put it. This is the sometimes elusive state where both mind and body become completely absorbed in the process of walking, an inspired union of energy and motivation; the body a well-oiled harmonious machine of muscle and sinew.

Mihaly Csikszentmihalyi, a Hungarian Professor of Psychology, is noted for his research into happiness and creativity and perhaps is best known for his ideas around this notion of 'flow' and particularly for his seminal work *Flow: The Psychology of Optimal Experience.*[15] His view is that people are happiest and most content when in a state of flow: a complete, joyful absorption within a particular activity to the point that nothing else seems to matter, a complete sense of immersion. Csikszentmihalyi spoke about the notion of the ego falling away; that when in a state of flow every action, thought and feeling follows from the previous one – a little like the intuitive, creative process of musical improvisation. In this sense one's whole being is involved. This is not a new concept of course, but simply reframed within the context of contemporary psychology. The idea of letting go of one's ego – of a state of mindfulness perhaps – has for thousands of years been practised by devotees of Buddhism, Taoism and other similar Eastern religions (and interestingly is now being integrated into ideas around both therapy and emerging neuroscience).

As a jazz pianist myself, I can certainly relate to the idea of being lost within the music, just ask Nicky who says that I have a marvellous but intensely irritating capacity for selective hearing when playing the piano. But I think this is also interesting in relation to physical activity. Runners often talk of entering the flow state – the blissful point at which they are more or less running on automatic – the state wherein I think running becomes quite addictive (on top of all the endorphins that is). I have no doubt this is the case for walking, especially long distance hiking, as it must be for all manner of activities within which we can become absorbed. Within therapy, we sometimes talk about dissociative states, a sense of disembodiment or zoning out, in which time passes almost unconsciously. Think of the times when you suddenly realise you have driven the last few miles without consciously being aware of it. At one end of the continuum this is a psychological defence mechanism to protect the mind from overwhelming traumatic distress. At the other end of this continuum it may have a pleasant daydreamy quality about it, the mind drifting off into all sorts of unconscious or semi-conscious nooks and crannies. When I am running, the best place to get to is that point when I drift off and lose all awareness of the fact that I am actually running. And there is also, with both running and walking, a very spontaneous, creative process at play when in this flow state and I know that when I am searching for inspiration for perhaps a lecture or a piece of writing, more often than not it comes to me when I am running or out walking.

So walking is a creative process. As Friedrich Nietzsche, the existential German philosopher once said, 'All truly great thoughts are conceived while walking', and indeed it is said that he found inspiration for the third part of his novel of eternal recurrence, *Thus Spoke Zarathustra,* while hiking the steep trail from the French Riviera up the rugged corniche to the precariously perched, medieval village of Eze. The story has it that Nietzsche climbed the 500 metres up to the village every day in the hallucinatory heat of the summer, composing the words in his head as he walked. The route, once a goat path and mule track is now officially known as Le Chemin de Nietzsche – Nietzsche's path. A deeply troubled man, his ideas were at different stages appropriated by both Zionists and fascists and, bizarrely, his favourite walking stick was given by his sister Elizabeth as a gift to Hitler in 1934. A poor fate for a great stick, that's what I say. Much good it did him... Hitler that is. Interestingly, the idea of the mountaineer was something of a recurrent motif for Nietzsche. 'The discipline of suffering – of great suffering,' he wrote, 'know ye not that it is only this discipline that has produced all the elevation of humanity hitherto... this hardness is requisite for every mountain-climber.'[16]

But the point I make – the point Nietzsche makes (amongst all the religion, morality and assorted nihilism) is that it was more often than not the process of walking that aided his creative thinking. Similarly, Henry David Thoreau, the American philosopher wrote: '... the moment my legs begin to move, my thoughts begin to flow', and Charles Darwin famously constructed a sand-walk circuit at Down House that he called his 'thinking path' around which he would walk several times a day as he mused upon his ideas of evolution by natural selection. The recuperative, restorative and creative powers of walking in nature are well accepted and deeply embedded in our culture, certainly in terms of literature and philosophy. (Take Thoreau's *Walden* for example, in which Thoreau details his experience of living in a cabin in the woods for two years and immersing himself – physically, emotionally and psychologically – in the natural world around him.) Now science and medicine are (increasingly) supporting the fact that walking, exercise and feeling connected to the natural world are not just healthy but evidenced-based alternatives to pharmaceuticals when it comes to the treatment of both physical and mental health. I know what I would rather be prescribed.

A couple of years ago researchers at Heriot-Watt University and the University of Edinburgh, using mobile electroencephalography (EEG) technology, attached portable EEGs to the scalps of 12 healthy young adults and sent them out for a walk in the park. EEGs record the electrical activity of the brain and the electrodes, hidden unobtrusively beneath an ordinary looking fabric cap, sent the wave readings wirelessly to a laptop carried in a backpack by each volunteer. The purpose of the study was to take a look at that condition of modern life that we are all too familiar with – brain fatigue, the point where our brains become overwhelmed by the constant noise, input and frantic demands of urban life. The researchers were focusing particularly upon urban life, but I think we can perhaps put 'brain fatigue' under the general umbrella of modern life. Urban or rural, we all tend to bathe in the nascent perma-glow of technology these days. But anyway, what was interesting about this study was that as soon as the walkers moved out of the city and into the relative tranquillity of the leafy park their EEG readings became calmer, quieter and more 'meditative'. Of course we know this instinctively, it's not rocket science. Being physically active in green open spaces is good for you and it feels good.

Part of my work as a child therapist has been about understanding the neurological impact of trauma and as I say, the world of therapy is becoming increasingly informed by neuroscience. It's the Next Big Thing. But it is exciting, especially because there is so much about the brain that we simply don't know or understand; it remains uncharted territory. Therapy aside, we are also learning more about the relationship between exercise and the brain and the impact our immediate environment has upon the quality of our lives, as the Edinburgh study suggests. Exercise affects the brain through a variety of mechanisms, including the release of all those lovely neurotransmitters, the brain's natural opioids that make us feel so good. These 'feel good' endorphins can be produced through a number of activities: sex, excitement, fear and even a good curry. There is a man at my gym (I call him 'the orgasmatron') who clearly suffers from something of an endorphin overdose when he exercises. It's like the infamous restaurant scene from *When Harry Met Sally*, although not quite as amusing. But anyway, where does all this rambling leave us? Walking is good. Mountains are good. Walking in the mountains is perfect. Get out there and do it.

Today, Ricardo – king of the TMB – seems to be in the zone and is charging up the hill like a man possessed, leaving the rest of us for dust. The path is mainly good, a well-worn track of earth and stone that winds ever upwards, but in some places it is eroded into narrow, furrowed channels of earth where it can be hard to find your footing. This section is less than 500 metres, but it is a tough, strenuous climb with the path twisting and turning its way upwards to compensate for the steepness of the slope. Ricardo is on fire today, and Claire and Julie are not far behind him. Rupert is struggling a bit and feeling the extra weight of his rucksack and he stops for an early lunch while the rest of us push on to the top. I wonder where he has got to for a moment and hope he is OK, but looking back I can see his tiny figure in the distance. It is important to find one's own pace on these strenuous climbs and to listen to what your body is saying, so I can understand Rupert's need to take it easy and walk in his own time and to rest when he needs to.

Eventually, after about two hours hard slog from the Rifugio Elena we haul ourselves up the last of the slope to the Grand Col Ferret, which sits at 2,537 metres. Exhausted from the climb, we throw down our rucksacks and collapse onto the soft grass and take a much needed lunch break. A number of other walkers are doing the same and this is one of those places that demand a moment of contemplation. This is, after all, the point where we leave Italy behind and is another significant stage of our journey. Ahead of us lies Switzerland, the path dropping away through a terrain of grass and rock to a backdrop of the Swiss Alps, and looking north-eastwards it is the snowy peak of Grand Combin that catches the eye. At 4,314 metres this is one of the highest peaks in the Alps and is a large glaciated massif made up of a series of peaks, three of which are over 4,000 metres.

The Grand Col Ferret stands on a ridge running northwest in the direction of Mont Dolent and looking back in the direction that we have walked there is a stunning view down through the Val Ferret and in the far distance, the Col de la Seigne, over which we passed all of three days ago. Standing here, looking back in time, it feels extraordinary to think that we have walked all this way, the entire length of this beautiful valley that cuts a dramatic swathe through the mountains, scooped out by one of my friends from above, the sky giants. While the others sit and finish their lunch I follow the ridge up a short way to try and absorb as much of this place as possible. Directly south from here lies Montreux and Lake Geneva, only about 60 kilometres away but it may as well be 600, reduced as we are to relative insignificance. To my left the jagged wall of the Grandes Jurasses still stands proud, but this will be the last we see of it for a while as we descend into the Swiss side of Val Ferret. The continuing sunshine leads us to linger here for much longer than we might do ordinarily and we sit and chat and eat for a while, just happy to be in the moment. We spend quite a lot of time comparing the weight of our rucksacks and we are all collectively gobsmacked by the weight of Rupert's. It is easily the weight of both Claire's and Julie's combined and it is little wonder that he has been struggling up the hills.

Eventually we move on, hauling on our respective rucksacks – helping Rupert with his – and head down on the track that leads through the grassy slopes, dimpled and pockmarked with little hollows and craters of earth and stone that will before long once again be topped with the fresh snows of winter. The path here is easy going, wide and well trodden and it snakes away into the distance, dotted with the tiny figures of other walkers further down the line. After about ten minutes our trusty pocket guide Mr Reynolds suggests that we take a short diversion up an unmarked path that apparently leads to a particularly fine outlook. It is good advice and we do indeed get a wonderful view down through almost the entire length of the Swiss Val Ferret towards La Fouly, today's destination. Blimey, it looks a bloody long way. Back on the path we weave our way down the right hand flank of the hillside through open, broad meadows of grass and after about another 45 minutes we are slightly taken aback by the unexpected sight of a couple of attractive yurts, sitting invitingly in an open terrace area in front of what looks like a farm building. We stop to have a little look around and see that they have tables set outside and are serving drinks, so we stop for a quick beer.

We soon realise this is the Gîte Alpage de La Peule and was one of the places that we and many other people had tried to call this morning from the Rifugio Bonatti in the frenzied search for accommodation. No one could get through so the general assumption was that it was either closed or full. Well, it certainly isn't closed and although probably a long shot I wander inside to see if they have any spare beds for the night as an alternative to the expensive room we have booked in the hotel at La Fouly. With a nod of the head a kindly woman leads me through to a lovely, new looking dormitory area and gestures to the fifty or so empty beds. *'Votre choix,'* she says with a smile. Your choice. It turns out the place is completely empty – the whole 58-bed dorm is totally uninhabited by other hikers. Outside, you can also choose to sleep in the straw of the two yurts and there are free hot showers as well as home-made food in the evening and plenty of wine and beer on offer. I go back and find the others who are still sitting outside finishing their drinks. 'Well,' I say, 'I think they can just about squeeze us in.' This place is a real gem and Richard, Rupert and I decide to stay for the night, even though it is only mid-afternoon and it will add another two hours onto tomorrow's stage. The girls are on a slightly different schedule, having arranged to meet Claire's boyfriend who is coming out to join her, so they decide to press on for La Fouly and then Orsières. It has been great walking with them and we pledge, marmot-like, to meet up further along the trail or perhaps at Chamonix.

The gîte is a working dairy farm and in the surrounding fields healthy looking brown cows snuffle in the grass chewing noisily, the enormous decorated bells round their huge necks chiming pleasantly every now and again as they flick their heads to dislodge the irritating flies that buzz around their ears. The dormitory is clearly a recently converted barn, and the two yurts look like they have a family in residence, the young children playing happily outside. The terraced area beyond the main dormitory building is a masterpiece of simplicity, a wood-chipped bordered area dotted with seating and interesting curios and artefacts: an old pair of walking boots used as a flowerpot, a disused but lovely ceramic stove and odd bits of furniture. Every object is simple but perfectly considered and there is a certain 'still-life' quality to it all that suggests someone here has both good taste and a very good eye. It is a rural refuge offering agricultural tourism; producing and selling their own milk from the cows; and making raclette, large wheel-shaped cheeses that are stored in the specific temperature and humidity of the underground cellar. Apparently, they also give tours of the milking parlour and cheese cellar and show people around the estate as well as providing lunches, teas and evening meals. And the setting is perfect, the gîte looking out across the valley with sweeping views in every direction. All in all it is hard to beat and it feels like we have accidently stumbled across a little bit of perfection up here in the Swiss Alps and what's more, we have it all to ourselves.

We make the most of the extra time we have by firstly having another beer and then washing out some clothes and hanging them up in the late afternoon sun to dry. In the dormitory we then make an exaggerated show of trying to choose where to sleep, testing out several bunks for comfort, just because we can. After our previous nights of forced intimacy we choose beds away from each other, happy to have a bit of personal space. A little later someone else turns up, a woman from Paris walking the TMB on her own. She seems friendly and we chat for a little while about our respective adventures along the way. For the remainder of the late afternoon and early evening, we simply hang out, enjoying the quiet, restful atmosphere of the place. It would seem that the owners, Nicolas and Sabine, began renovating the old farm buildings back in 2008 and the gîte was up and running a year later, which explains the fresh, new feel of the place, although this is tempered by the lovely original wooden beams that run through the building.

Later on we have dinner in the dining room, warmed by a log-burning stove around which we have draped our assorted wet socks in the hope that they will be dry for the morning. The yurt family are eating at the far end of the room and it seems they are either locals or friends of the owners. Dinner itself is a massive chunk of bread topped with ham and melted cheese, a kind of cross between a pizza and a croque-monsieur. It's delicious and nicely helped down with a carafe of red wine. We are then treated to the full Swiss raclette experience as a whole round cheese is brought out and attached to a huge and very ancient looking metal device that grips the cheese and then slowly turns it over a hot flame. A scraper then takes off the top layer of the cheese, forming a melted pile of cheese on the plate below, which is served with boiled potatoes and more bread. Again, it's delicious, although there is a part of me (most likely my arteries) that is aware we have just had a large meal of melted cheese and bread followed by, well, more melted cheese and bread with a few potatoes thrown in for good luck. We wait with baited breath for the final course, possibly cheese and biscuits or even cheesecake. Thankfully it is neither. Is fruit the Swiss antidote to cheese?

Seventeen

As I approach the age of 50 I circle it cautiously a little like Rupert and I are circling this mountain range, the inevitable portents of the ageing process become ever more apparent and at an ever more frightening rate of acceleration, both physically and emotionally. It is at this age that you begin to become aware that the people around you – friends, family, colleagues, acquaintances start getting ill, dying even. Cancer rears its ugly head and life becomes something of a lottery, an anxious game of spin the bottle in which you never know who is going to be next or, more to the point, why. There seems to be a tragic randomness about this that leads to all kinds of emotional re-evaluations, a shifting of perspectives and repositioning of priorities as we rethink what's important in our lives. Suddenly, we are confronted with not just the idea or possibility, but with the absolute *reality* of our own fragile mortality.

As I said earlier, I think I have strangely become more relaxed with this as I have got older, certainly more relaxed than I was in my angsty forties, but I am not sure if I quite believe anyone who says that they don't mind the idea of dying. Perhaps it is, as someone said, not death itself we fear most but the incompleteness of our lives, and I am sure that this holds true for most. We are all touched by the tragic stories of people who have died too soon and the realisation that life – as our parents always told us – is not fair. The truism of my mother's motto – live each day as if it were your last – comes to mind.

And our bodies, with a brutal honesty, like to remind us of the fact that at 50 we are beginning to get 'really quite old'. My knees are starting to complain about the constant battering they get from the running and walking. I have tinnitus from years of playing in loud bands and can't hear very well in noisy places. I have finally, after years of resistance, had to admit that my eyesight is deserting me and get used to wearing glasses. (The only plus is that you get to play Space Invaders at the opticians when your eyes are tested.) I forget things and my hair gave up the ghost years ago – or has just relocated to my ears. Recently, when going to the cinema with Nicky, the young girl at the ticket desk asked me if I was a senior – a whole ten years out! All in all, I am slowly but surely turning into Mr Magoo – shortsighted, hard of hearing and cantankerous.

So we can only do what we can and as well as walking and running I regularly go to a local gym in a continuous effort to keep the ravages of Old Father Time at bay. The gym is a place that encapsulates all sections of life, a microcosm of society writ large. There are the lithe young pretty things, who effortlessly sweep around the place as if their feet barely touch the ground – a depressing vision of what once was – and the parents who are herding their startled looking brood into ballet and tae kwon do classes. Then there are the body fascists, the sculpted muscle-bound narcissists absorbed in the liquid pool of their own image. These are the closet angry people, building and shaping a well-honed carapace around their fragile vulnerabilities. You have the personal trainers of course: cool, fit and just slightly superior compared to us mere mortals. If you go on a weekday morning, you might catch the Saga-plus brigade, admirably working out in the late autumn of their lives – mostly under doctors' orders I would hazard a guess, absorbing the liquid pool of their own incontinence.

And then you have people like me – men who are slipping off the edge of the 40–49 demographic and turning into reluctant 'veterans' – the middle-aged, middle-classed and middle-sized who are all having to cope with the reality of going to the toilet more than once in the night and making strange grunting noises every time we get up from a comfortable chair. I tend to keep to myself at the gym – the lone worker – but I am aware that in the male changing room there are only two main topics of conversation: the declining physical condition of the human body and home improvements. Seriously, this is all men seem to talk about. Clearly there is a corresponding relationship between the two. As another knee finally gives out, let's go one last yard on that loft extension. As that pesky lower lumbar disc pops out once again, let's make the final push on that conservatory. This is, I think, the late middle-aged male version of nest-building, feathering the house with the accoutrements of old age in preparation for the time when you know you are not going to be able to get out any more. I also wonder where these people get all their money. The only feathering I am going to be doing is perhaps another antique brass car horn to attach to the bannisters, and that's not going to help me a great deal when I am doddering around in my dotage. (Well, I suppose I could wear it around my neck – a retro personal alarm.) Not long ago I was at the gym and passed by a class of middle-aged folks doing some form of yoga. 'Do you think I might be able to join your yoga class?' I asked the instructor. 'Hmm, how flexible are you?' she asked, looking me up and down. 'Well,' I said, 'I can make any evening apart from Tuesdays and Thursdays.'

Life is but a long distance hike. One sets out, fresh and eager with a veritable skip in the step and if you miss an occasional route marker, or through simple curiosity decide to take another path, it is of little consequence – you will soon bounce back and find yourself once again on the main trail. Then you settle into the main phase of the hike – the part that takes in most of the territory, where most of the hard work is done. This is the 'head down' and 'long hard slog' part of the walk. Wrong turnings can be hazardous and as tiredness sets in little trips and stumbles are inevitable. Indeed, long periods of being lost in the upland mists may be experienced.

And then it is the last leg, the bit that you have been looking forward to but realise, once there, that the view is not nearly as great as it was back at the top. This is the stage where I and my fellow fifty-something, gym-going 'life-hikers' find ourselves, looking forlornly back over our shoulders and hoping desperately, but quietly, that we might stumble across a magic switch that puts our treadmill in reverse. But unlike the mid-life existential angst of the forty-somethings, there is perhaps a greater sense of acceptance rather than denial at this stage in our journey, that we have almost arrived at life's final *rifugio* and that actually it's... well, alright really (as long as it isn't anything like the Elisabetta of course!).

Walking around Mont Blanc symbolises so many things: life, childhood, family, hopes, dreams and regrets. (You can take the walker out of the therapist, but you bloody well can't take the therapist out of the walker!) But perhaps also, as Freud might say, sometimes a mountain is just a mountain – the thing is the thing. To be fair, he probably wouldn't say that. He would say it is a symbol of my repressed unconscious; of sex, death, anxiety or perhaps the maternal breast rising out of great Mother Earth (actually that's more likely to be Klein)... or my dead father perhaps, whatever. I don't know whether I am walking towards something or away from something, maybe neither and both, but the fact is that by the sheer virtue of spending several hours every day walking in some of the most spectacular scenery imaginable, one can't help but reflect upon life as you go; it's the nature of the beast, so to speak.

Funnily enough, as well as walking around Mont Blanc, I did once drive underneath it with Nicky about 25 years ago. We were on holiday, driving aimlessly around France in her old white Mini and decided that we would pop over into Italy via the Mont Blanc tunnel. For a while this was the world's longest road tunnel and was a considerable feat of engineering. Begun in 1957 and completed in 1965 the tunnel is 12 kilometres long and took 711 tonnes of explosives to blast 555,000 cubic metres of rock, eventually linking Chamonix on the French side with Courmayeur in the Italian Aosta valley. Prior to the tunnel's opening, Courmayeur was a much more isolated, quieter and more underdeveloped town than Chamonix and perhaps in this sense the tunnel was something of a mixed blessing for the residents. Certainly it raised Courmayeur's economic profile and made it a much livelier town, but I would guess that it was not to the liking of everyone – these things invariably never are. Anyway, I recall that upon emerging at the Italian end of the tunnel the brakes on Nicky's Mini suddenly and completely gave out and we had to crawl our way downhill by virtue of the gears and handbrake, eventually somehow making it back into France to get them repaired. So I will have been round and under this mountain, which only leaves going over the top.

On the French side of the Mont Blanc tunnel stands a commemorative plaque in honour of the 39 people who died in the horrific fire of 24 March 1999. It was a terrible disaster in unimaginable circumstances and it is hard now to think about the tunnel without reflecting upon the people who lost their lives on that tragic day. One of those who died was Pierlucio Tinazzi, an Italian security guard, whose job it was to patrol the tunnel on his BMW motorbike, keeping the traffic flowing and providing assistance to any drivers who might need it. On the morning of the fire, Tinazzi was on a rest break and as soon as he was notified of the developing emergency he without hesitation grabbed his breathing equipment and headed into the tunnel. Passing people trying to get out, Tinazzi advised them to stay low, remain close to the side of the tunnel walls and keep moving and then, despite the risk to his personal safety, he continued on to the most dangerous and hottest part of the fire. Searching the wreckage for people who were trapped or had succumbed to the heat and smoke of the fire, Tinazzi hauled survivors onto the back of his bike and drove them out to the French side of the tunnel, shuttling back and forth bringing out victim after victim. On his fifth trip back into the blazing tunnel, Tinazzi found a French truck driver, Maurice

Lebras, unconscious and too big to get onto the back of his bike. Refusing to leave him, Tinazzi hauled him into one of the tiny fire refuges built into the sides of the tunnel. However, the refuge was unable to withstand the intense heat of the fire that blazed for over 50 hours. Tinazzi's BMW bike was later discovered melted into the pavement just metres from the door to the refuge. His bravery that day saved the lives of at least ten people and he was posthumously awarded the Gold Medal of Civil Valour by the Italian Government. Pierlucio Tinazzi, a true hero.

After a peaceful night at the Gîte Alpage de La Peule, we have a good breakfast of coffee, toasted home-made bread, jam and… cheese. This is possibly the cheesiest place I have known. The folks running the place are delightful and when we realise that we are now in Switzerland and have no Swiss francs, they allow us to pay in euros. After showering and packing I sit in a chair on the terrace and watch the sun rise above the peak of La Tsavre (also known as Col Ferret) almost directly opposite me across the valley. It has been a good stop here, unexpected but very welcome, and Rupert and I could imagine us at another time spending a day or two here, simply lingering in the peaceful ambience of the place, absorbing its rich tranquillity. The idea of sleeping on a bed of straw in the yurt is also very appealing. The plan today is to head down into La Fouly where we can find a bank or cashpoint to try and get our hands on some Swiss francs and then head on to Champex. Because we stopped earlier than we planned yesterday we have left ourselves with a longish seven-hour hike of over 20 kilometres, and although it is apparently one of the easier sections of the TMB, certainly one of the flatter stages, we have learned from experience that 'easier' is very much relative in these parts.

So suited and booted, we leave the serenity of La Peule and head off under blue skies and a glorious morning sun, the weather once again surpassing all expectations. I heard once about someone who walked the entire TMB and never actually once set eyes on Mont Blanc due to the persistent rain, mist and cloud cover, so we are thankful for the incredible weather we have had so far.

The first stage of the walk takes us down on a path that traverses the side of the valley, gradually dropping down towards the little band of buildings at Ferret, which sits on the floor of the valley. The trail is a well established mud and stone track gently winding its way through a covering of bilberry shrubs; open pastures of green, dotted with white flowers; and the ever present rowanberry trees – the red clusters of berries hanging with crystalline clarity against the backdrop of a blue sky dappled with wispy white strands of cirrus cloud. The morning is alive with vibrant colour, vivid in the thin alpine air and set against the constant stone-grey and snow-white of the mountains, which seem to occasionally shimmer in the distance as if they are but a mirage, a fanciful figment of my fevered imagination. It is a truly sensual experience, its felt immediacy so intense that paradoxically it almost feels unreal – a hallucination of sorts. A sharp, glacier-cooled morning breeze whips down from the mountain tops and sends a ruffled shiver through the trees as we pass through wooded sections of hillside and in the shade it feels cold until we emerge again into the warmth of the morning sun. Now and again as the trail takes us around a contoured spur in the hillside, we get spectacular views of the peaks of Mont Dolent and Tour Noir to the northwest, their attendant glaciers sending icy fingers downwards as if reaching for the valley floor.

The sharp steeple of Mont Dolent rises to 3,829 metres and is a peak first climbed by Edward Whymper in 1864 – Whymper being best known for his ascent of the Matterhorn a year later. In his book *Scrambles Amongst the Alps in the Years 1860–69*, Whymper says of Dolent:

> *This was a miniature ascent. It contained a little of everything. First we went up to the Col Ferret and had a little grind over shaly banks; then there was a little walk over grass; then a little tramp over a moraine (which, strange to say, gave a pleasant path); then a little zigzagging over the snow-covered glacier of Mont Dolent. Then there was a little bergschrund; then a little wall of snow, which we mounted by the side of a little buttress; and when we struck the ridge descending south-east from the summit, we found a little arête of snow leading to the highest point. The summit itself was little—very small indeed: it was the loveliest little cone of snow that was ever piled up on a mountain-top; so soft, so pure, it seemed a crime to defile it. It was a miniature Jungfrau, a toy summit: you could cover it with the hand.*[20]

Whymper makes it all sound very effortless and, well, very little. It doesn't look little from here.

Mont Dolent aside, the walking for us so far is pretty effortless and I make the most of it by hitting Rupert and Ricardo with another round of bad jokes and I am slightly surprised myself by the depths of the endless well of poor humour from which I draw. Ricardo is particularly amused by my 'little otter' joke and he chuckles away to himself for at least two kilometres. He says that while he is useless at telling jokes he is determined to remember this one and we suggest that he could impress Claire and Julie with it when we meet up with them in a few days. So Rupert and I spend the next half an hour coaching Ricardo in the art of joke telling – getting him to repeat the joke again and again until he has it just about right. He is excited by his challenge, but Rupert and I are not sure what we have unleashed upon the world. Will we be able to get the genie back in the bottle? We soon drop down to the valley floor and pass through Ferret, a hamlet that sits dwarfed by the mountains above and we pass a striking chapel, dating back to 1707 and dedicated to Our Lady of the Snows, with columned supports and arched windows with matching wooden arched shutters.

The path then rises slightly before descending through a meadow into a wooded area and over a wooden bridge across the river. This is the approach to La Fouly and the hillside around the village is dotted with pristine wooden chalets and there is no mistaking the fact that we are now in Switzerland. The difference, now that we approach habitation, is really quite striking. Everything is very pretty, incredibly neat, clean and ordered without a thing out of place. Even the allotments are perfect, with uniformly straight, regimented rows of pristine leeks standing proudly to attention. There is a uniquely Swiss quality to the scene and it is hard not to stereotype, although the reality lends credence to that very self-same stereotype. Sitting at the foot of Mont Dolent and the Tour Noir, it is easy to see how La Fouly functions as a perfect base for walking in the summer and skiing in the winter, as well as having a regular stream of hikers on the TMB passing through. It's a little place and we find the tourist office and ask the friendly girl inside to help us ring ahead to book accommodation in Champex, where we plan to stay tonight. Many places are booked up and the girl is very patient as we ask her to work through the list of numbers in our guidebook until eventually she secures some beds in a refuge just outside Champex. I feel a little sorry for this girl – she must have to deal with a succession of TMB hikers all asking her the same thing.

Eighteen

La Fouly is one of 21 villages that form the commune of Orsières, which covers a total of 165 square kilometres, bordering both Italy and France. The town of Orsières itself, just a short hop from La Fouly up the river Dranse, benefited historically from sitting astride the junction of two Alpine valleys, Val Ferret – which yesterday brought us from Italy into Switzerland – and the Great St Bernard Pass that runs south-eastwards from Orsières through the Valais Alps down to the border where it crosses into the Aosta valley in Italy. This famous crossing, which as the crow flies lies just a short distance south of where we are now at La Fouly, is the most ancient pass through the Western Alps and has certainly seen the passing of history; once being a Celtic track, then a Roman road and then later part of the Savoy royal route. The route traversed the Dranse in Orsières and it was the strategic importance of the bridge at this crossing point, permanently guarded in Charlemagne's time by a garrison of soldiers based in a glacial hollow, which laid the foundations for the village that gradually grew up around it.

The earliest recorded passage across the Great St Bernard Pass, or Mont Joux as it was once known, is that of the Celtic tribes in the invasion of Italy in 390 BC and despite an initial failed attempt by Julius Caesar in 57 BC to seize control of the pass the Romans finally made it their own through the efforts of Caesar's adopted son, Augustus. A large *castra* – a military defensive position – was subsequently established by Augustus at the foot of the pass, which latterly became Aosta. In 1800 Napoleon famously crossed the pass with 40,000 men carrying heavy artillery as they prepared to battle the Austrian army that had laid siege to French-occupied Genoa.

It must have been quite a sight as over the course of several days Napoleon's army crossed the pass in single file at a rate of 6,000 men a day, each carrying up to 31 kilograms of disassembled arms, ammunition, gun carriages and artillery; the cannons in hollowed out pine logs dragged up through the snow by mules and men. Military bands played motivational martial music along the way to keep the men's spirits up, with drum rolls to keep them alert during the particularly difficult sections, and the story goes that at the monastery at the top of the pass monks handed each passing soldier a slice of rye bread with cheese and two glasses of wine, although it may well have been more by the sound of it. It is recorded that during the crossing 21,724 bottles of wine were consumed plus a tonne and a half of cheese and over 800 kilograms of meat, running up a rather grand bill with the monastery and hospice of Fr.40,000, a bill that Napoleon summarily dismissed with a casual wave of his hand. Fifty years later, the monks received Fr.18,500 in part payment and had to wait until May 1984 when the French president François Mitterrand made a token gesture of settling the long-outstanding account. The rather large amount of wine consumed by Napoleon's men may account for the fact that when they reached the other side of the pass, where the snow had become densely packed by the relentless marching, they gave up walking and took to sliding down on their backsides, including Napoleon himself – the last man down. It all sounds rather like a particularly extreme cross-country half marathon over Box Hill in Surrey that I took part in recently, minus the heavy artillery.

Of all the many monasteries scattered across the Alps, the one at Great St Bernard that sits at 2,469 metres atop the desolate summit of this historical mountain pass is probably one of the most renowned. The community was first established around one thousand years ago by King Canute of Denmark who despaired at the numerous Saracen attacks upon the many European traders who used the pass and Bernard, Archdeacon of Aosta, was asked to found a monastery and hospice at the summit of the pass to protect and shelter travellers as they travelled through. Having already founded a hospice at the top of the Petit St Bernard Pass, this one was consequently named the Great St Bernard and so the Archdeacon Bernard spent his years ministering to his peoples in the mountains. He became St Bernard in 1923 when he was named patron saint of the Alps by Pope Pius XI. The monastery grew increasingly wealthy as a result of the many gifts bequeathed by grateful travellers who had been afforded safe haven as they crossed the pass, and by 1177 it owned land all over Europe, including even a plot in Hornchurch, Essex, where St Andrews Church stands now. Clearly, the monastery was well positioned, sitting for many hundreds of years at the head of what was one of the busiest routes across the Alps, although now of course it has seen quieter times since the construction of the road tunnel running underneath it as well as the Mont Blanc tunnel and the Simplon rail tunnel. But the monastery is still active, housing a tiny community of monks in two buildings that date back to 1560 and 1898.

Many notable people have visited the monastery and crossed the pass during times gone by, with mixed impressions. The aforementioned alpinist Edward Whymper described his visit with about as much prosaic pragmatism as he did his summiting of Mont Dolent. He said of his trip to the pass, 'I have not had any day so devoid of interest and barren of incident, neither have I walked over so uninteresting a road' and dismissed the scenery at Great St Bernard as 'commonplace'. Clearly Whymper was used to something a little more dramatic – perhaps swinging by his fingers from an icy precipice. Someone a little more taken by the experience was Charles Dickens, who in a letter to his friend and later biographer John Forster described the summit at the pass as:

> *... a great hollow on the top of a range of dreadful mountains, fenced in by riven rocks of every shape and colour: and in the midst, a black lake, with phantom clouds perpetually stalking over it. Peaks, and points, and plains of eternal ice and snow, bounding the view, and shutting out the world on every side: the lake reflecting nothing: and no human figure in the scene.*[1]

In fact, in *Little Dorrit*, Dickens draws parallels between the bleakness of the monastery on the top of the Great St Bernard Pass and the Marshalsea prison in London, the debtors' prison in which Little Dorrit's poor father William (the 'father of Marshalsea') was incarcerated for much of his life. The handful of monks in the Great St Bernard monastery, while not prisoners, do exist in self-imposed exile and it must be a tough, hard life during the winter months when the visitors have disappeared and where the heavy snows do, in a sense, hold them captive. Although Dickens' portrayal of the Swiss mountains is rather bleak – Dickensian one could say – he was by all accounts rather enchanted by the Alps and in his later years had a mock Swiss chalet constructed in the garden of his house in Rochester, Kent. Instead of snow-capped mountains his view from the chalet window would have been of Kentish fields, steams and sail boats wending their way down along the Thames estuary.

So, over the course of history any number of travellers, merchants, pilgrims, kings, emperors and even popes have passed through this great Alpine gateway. Hannibal's famous crossing of the Alps with his elephants in 217 BC is also often associated with the Great St Bernard Pass, although there is in fact little evidence to support this (there being much debate about several possible routes that Hannibal may have taken). And of course we have to mention one of the defining icons of the Great St Bernard Pass, our large and hairy friend the St Bernard dog, which was employed at the aforementioned monastery as far back as 1690 to sniff out poor souls lost in the snow and to revive them with a legendary little cask of brandy tied around their collars. The hospice at the top of the pass still keeps a kennel for the dogs and fifteen or so pure bred St Bernards are born up there every year.

Dogs aside, the next challenge we have to face is getting our hands on some Swiss francs and so we find a cashpoint and then spend an inordinate amount of time trying to work out how many francs we might need for what looks like two nights in Switzerland. It sounds a reasonably straightforward task, but somehow becomes unreasonably complicated as we try to figure out the relative exchange rates, knowing also that the refuges will take cash only and that we will also need enough euros to get us through to the next place where we can get money. Perhaps it is a result of being 'off-grid' for so long, but our brains have seized up and we seem to have lost the capacity for logical thought. One result of this is that Ricardo, who is down to his last few euros, withdraws about £200 in Swiss francs and then quickly realises that he will have to change most of it into euros before crossing back into France, at a significant loss due to the exchange rate. The instant panic this induces is mildly amusing and seeing him suddenly so loaded with francs that he can't spend inevitably results in him being renamed 'Swiss Ricardo'. Tooled up and minted, he is rapidly turning into some kind of Alpine gangster figure and it wouldn't surprise us if he uses his new-found wealth to start up some kind of TMB protection racket or money laundering service.

We stop for a brief coffee and also find a shop to stock up on bread, cheese and fruit for lunch and then continue on our way through the village where the trail then drops to the left through larch trees and crosses the river over some kind of sluice system. From here the route of the TMB follows the left hand side of the river along the floor of the valley and the going is very pleasant. In the September sun the deciduous larch conifers exude a rich autumnal glow, their needles yet to fall and the easy path, never straying far from the course of the river, weaves in and out of the trees as we pass day trippers and picnickers making the most of the clement conditions. After a short while we pass a large area of exposed rock face rising several hundred metres to our left, the Dalle de l'Amône slab, and it looks like a few rock climbers are preparing to get in some practice.

At one stage we pass a lovely looking little stone shelter by the side of the river and it occurs to me that it would make the perfect dwelling for a hermit, one of my many 'escape' fantasies. The house also reminds me of the Greek myth of Procrustes, the evil smithy – the 'stretcher' – and son of Poseidon who waylaid passing travellers with offers to rest up for the night. The story goes that if the hapless traveller didn't fit his iron bed Procrustes would either cut off the legs or stretch the body of the traveller to ensure that he or she did fit. Nobody ever fitted the bed because Procrustes secretly had two – the sneaky devil. Still, he got his comeuppance when Theseus finally killed him by fitting him to his own bed, one way or another. I tell Rupert and Ricardo the story of Procrustes and my fantasy about living in this little house by the river, but they are not convinced that it would be one of my wisest career moves and Ricardo is too occupied with practising his otter joke to concern himself with Greek mythology.

We continue following the left bank of the river before picking up height slightly and traversing the side of the valley that drops down, often steeply, to our right. At one stage we meet a friendly group of Brits, northerners from Yorkshire who seem to be part of a social walking group that have come out here together to do the TMB. We stop and chat for a while, exchange notes on our experiences of the last couple of days; they are nice folks and I like these brief encounters on the trail. The path then zigzags a little before veering to the right and leading us along an unusual ridge formation, the Crête de Saleina, created by the lateral moraine of the Saleina glacier, which has long retreated back towards the peaks that rise unseen beyond the trees to our left. The path is like a narrow causeway, the wooded slopes falling away on either side and it is both a striking example and dramatic reminder of the rich glacial history of the landscape; carved, shaped, sculpted and ultimately abandoned by the mighty rivers of ice that once pushed and sliced their way through the earth and rock.

At the base of the moraine we cross the tributary river, the Reuse de Saleina, and the trail opens up to some broad sloped meadows to the little hamlet of Praz-de-Fort, with its picturesque chalets of stone and timbered brown roofs, all shutters and flowers. Many of the houses have been converted from the old larch wood raccards – the traditional granaries found in the Valais region of the Swiss Alps that are built off the ground, supported by wooden stilts. Traditionally, a circular stone slab, forming an overhang, would be placed between each stilt and the granary building as a way of preventing hungry rodents from getting to the precious grain. They are lovely structures, an aesthetic feast of rich wood and functional simplicity and like my little hermit's house that we passed earlier, I fantasise about living in a place like this.

The name Praz-de-Fort means 'field of ovens' and has its origins in the lime that was fired in open kilns in the fields, and the village itself is known for its unusual maypole tradition, going back to 1833 and resurrected in 1944, when each year on the first Sunday in May a huge spruce tree – the *Mai* – is carried into the village square and hoisted into place by a team of brave villagers. By all accounts it is a delicate operation, the *Mai* having both fallen and been struck by lightning in times gone by. But the spring celebrations are long distant now and as we wander through the quiet lanes we can see that the barns and houses are all stocked with cut wood in preparation for the long winter ahead, meticulously stacked with not a log out of place. The heart of the village centres on a grouping of narrow lanes between a cluster of houses and it is all very lovely and very Swiss in a way that does not seem quite real, a little like a Lego Playmobil version of a Swiss village. The illusion is added to by the fact that it is so quiet, not a soul in sight, and we guess that many of these houses are holiday chalets and used more in the summer and winter seasons, or else everyone is indoors having their lunch, which is also very likely. The wall of the moraine maintains the ghost of a glacial presence over the village and out of the trees now, we can see the striking peak of the Petit Clocher du Portalet pointing a spiky finger towards the heavens.

We pause for a breather in the village and in the thickening heat of the midday sun are grateful for the opportunity to refill our water bottles from a fountain of cool fresh mountain water. While the walking has been reasonably undemanding, certainly compared to what we have become accustomed to over previous days, it is still tiring and it is nice to rest up for a moment. I also take the opportunity to wash out my hat which has developed a nice crusty ring of salty sweat around the rim and is in much need of attention. Then it is onwards and we leave the Field of Ovens on the single main road that snakes its way along the valley floor before crossing the river again where the path takes us across a lovely area of open meadows towards the hamlet of Les Arlaches. The pastures are a painterly sprinkling of September colour, the verdant uncut grass laced with the delicate blues and yellows of gentian flowers, as well as the fluffy cottonwool white of the linaigrettes that dot both the meadows and hedgerows of the path. Once again, there is a somewhat cultivated and uniquely Swiss quality of perfection about the scene, the flowers looking like they have each been individually hand-placed with the utmost care, as if by some force of intelligent design.

Before long we pass through the little hamlet of Les Arlaches, the colourful restored houses of the central square giving the place a certain charm and there are a couple of fine examples of granary raccards dating back to 1773 and 1779. This is an area rich in its farming traditions; the growing of wheat and the communal baking of bread, and in acknowledgment of this the villages of Les Arlaches, Praz-de-Fort and Issert are joined by the walking trail known as the Sentier du Blé – the Wheat Walk – that celebrates the old local farming practices of the region. In fact, as we walk through the village we can see Issert in the distance and as we meander by the inviting shade of an old tree we decide, in the spirit of the Wheat Walk, to stop and break some bread together and have a little lunch, having been on the go for a good five hours or so. It has been pleasant walking together, chatting good heartedly at times and walking in comfortable silence at others. Although really only days, it feels like we have been together for much longer, our merry band, and Ricardo has kept Rupert and myself amused with his various plans to try and offload his Swiss francs. He has also been in contact with Claire and Julie who are now a day ahead of us due to our unscheduled stop in La Peule and there are vague plans for us to try and meet up further along the trail.

Fortified and refreshed, we continue on our way and it is not long before we pass through the village of Issert that sits astride the main road through the Val Ferret and adjacent to the river Dranse, whose course we have been following from La Fouly, and which if we continued to follow would take us all the way to Lake Geneva where it deposits its icy mountain waters. Adorned with pots of red geraniums and hanging baskets that burst with a vibrant combination of purple, pink and white, the houses look old, solid and well cared for and the village has a charm to it, slightly diminished by the main road that rather brutally dissects it. Originally built upon granite erratics, huge boulders carried down and then abandoned by the retreating glaciers, we can see how the boulders have been incorporated into the foundations of some of the houses, like great cankerous growths that bulge out of the walls, an indication of just how old some of these houses truly are. At the upper end of the village on the left bank of the Dranse stands the old mill, restored to its former glory and apparently dating right back to 1633 and still capable of grinding rye, wheat and corn. Spotting a bar we are tempted to stop for a beer, but the day is quickly passing and we still have a couple of hours to go and a climb of around 500 metres before we reach Champex, so we decide to press on.

Leaving Issert behind us the trail veers away from the road and we cross a small stream where the path enters the forest and we begin the long climb up to Champex. Having been spoiled by the relative flat of the valley for most of the day the sudden increase in incline comes as something of shock to the system and while our trusty pocket guide, Mr Reynolds, describes this ascent as 'modest' it doesn't feel so, particularly in the oppressive heat of the late afternoon sun. However, the trees provide some welcome shade and the well worn path weaves its way up through the dense forest of pine, the relentless twists and turns compensating for the gradient. As we climb higher we pass several wood carvings of birds, squirrels and some very phallic looking toadstools and it turns out that this is the Sentier des Champignons, the Mushroom Trail, an educational route established by the local mycological society – a celebration of all things fungi. The path continues to weave its way through the trees, passing inviting benches and picnic tables that tempt us to stop and rest up for a while, and occasional spurs in the path offer us a grand eagle-eyed view of Orsières, a miniature town laid out upon the valley floor beneath us.

Eventually, a couple of hours after leaving Issert, we stagger exhausted out of the woods and enter the outskirts of Champex and the first building we lay eyes on is the charming looking Hôtel Belvédère. On impulse, lured perhaps by the suggestion of comfort, I call in on the Belvédère and ask if they have any rooms available and a charismatic, rather eccentric seeming woman that we take to be the owner says she does have one room available. The cost is really beyond our budget and we are already booked into a large 55-place refuge on the other side of town, but I am feeling a little rough with a developing case of 'dormitory throat' and the idea of a night of comfort is very appealing. Undecided, we head into the centre of the village to get a drink and consider our options.

Champex is a delightful place. Sitting at a height of 1,500 metres, the village is coiled around the still, blue-green waters of the central lake, which provides a glassy mirror to the surrounding peaks that jostle impatiently together as if they are crowding in to bathe in their reflected glory in the glacial water that lies cradled, bowl-like, within the tree covered folds and ridges of the mountains. The mountains themselves look as if they have been fashioned absent-mindedly by my ever present playful friends the sky giants upstairs. It is a small village with just a few hundred inhabitants and has a peaceful and enchanting quality about it. This aura is perhaps partly as a result of our climb up from the valley floor and the subsequent change in height giving Champex a rather unworldly feeling; a floating village like a castle in the sky. On the eastern side of the lake the road and footpath curl around with the contours of the water and take us towards the main heart of the village, while looking across to the other side we can see the pine carpeted hillside rolling down to the water's edge. A number of small wooden rowing boats lie clustered together, no doubt busy in the summer months, and a handful of fisherman dot the bank of the lake, quietly absorbed.

It was back in 1850 when Champex first came to prominence, thanks to a French historical writer by the name of Emile Bégin who had been touring the St Bernard region of Switzerland as part of his research into the life of Napoleon. Bégin's travels took him up to this little oasis in the mountains, surrounded by fir trees and green pastures grazed by cows and chamois, and in a book that he later had published, *Travels Picturesque to Switzerland*, he included a section on this place that he had been rather taken by, which put Champex on the tourist map. Gradually, as more and more walkers and alpinists discovered Champex, the wooden cattle sheds with their basic living accommodation above the cows slowly became inns offering a humble bed for the night, perhaps along with a simple meal of rye bread and cheese and a drop of the local wine.

Sometime later, in 1892, a chap called Daniel Crettex who was the son of an acclaimed mountaineer of the time, Maurice Crettex, built the first official hotel in Champex, the Hotel de la Poste, where apparently the guests could refresh themselves by bathing in milk if the fancy so took them. Things then picked up when the Simplon railway line was opened in 1906 and visitors could be brought to the village in horse-drawn carriages from Martigny station and with this sudden influx of people it was only a matter of time before Champex emerged as something of a resort, boasting nine hotels and all manner of alpine attractions. The railway line, and particularly the Simplon tunnel (which is in fact two single track tunnels built 20 years apart) was a major feat of turn of the century engineering and was the longest railway tunnel in the world until it was eclipsed by Japan's Daishimizu tunnel in 1982. During World War Two preparations were made on both the Swiss and Italian sides of the border for the Simplon tunnel to be detonated and bizarrely, the explosives from the Swiss end of the tunnel were not removed until 2001. So much for Swiss efficiency. Back in Italy, the Germans themselves planned to blow up the tunnel as part of their 1945 withdrawal, but thankfully were foiled by Italian partisans with a little help from two Swiss officials and some Austrian deserters. So all in all it was a good thing (particularly for Champex) that the tunnel survived.

Champex has also benefited from a number of philanthropists and benefactors, one of these being the Freudenberg family, a prominent German Jewish family that took refuge in the village during World War Two. The Freudenbergs took up residence on an island in the lake to protect themselves from the Nazis and were also responsible for helping many other Jewish families escape across the border from France. The island is now known as the Iles de Freudenberg, and as a thank you to the village for offering them sanctuary the family, between 1920 and 1993, constructed a network of paths and walkways above the lake – rambling trails through bilberries and rhododendrons and balcony viewing points high above the village that are still in use today. In 1952 a chairlift to the peak of La Breya was opened that subsequently led to the village also becoming a popular skiing resort in the winter, finally adding the snowy icing on the tourist cake and giving Champex a year-round appeal.

We find a pleasant cafe with tables skirting the water's edge and enjoy a cool beer that tastes so much better for the seven hours and almost 25 kilometres it has taken us to reach it. One of the lovely things about a long distance hike of this nature is that life becomes reduced to a series of simple pleasures that all become worth so much more due to the context in which they are taken. The first heavenly swig of ice cold lager, a dried apricot and a square of dark chocolate, an apple, a sip of water and a hot shower all become small things of magnificence that have been earned through the hard but pleasurable graft of walking; delightful parcels of perfection to replenish a body depleted of energy and to restore muscles, tense and tight from hours of repetitive strain. Today has not by any means been the hardest day, in fact on paper it has been the easiest, but it has been a long stretch and we all feel exhausted. It is about another 30 minutes' walk through the village up to the refuge and Rupert and I have pretty much decided that we are going to try and stay here in Champex for the night. My case of 'dormitory throat' is rapidly turning into a full-blown cold and I am starting to really feel quite rough. Meanwhile Ricardo is accosting anyone he can find in a bid to exchange his Swiss francs… unsuccessfully it would seem. We suggest he could at least get rid of some of them by buying another round of beers.

So from life's simple pleasures, Rupert and I decide to indulge in a night of comparative luxury and we go back and book ourselves into the Hôtel Belvédère. Ricardo is unimpressed, accusing us of 'wimping out' and questioning our credentials as hard-core hikers. He is right, but my philosophy on these kind of long distance treks has always been that one should be open to at least one night of comfort and comparative excess – if only to have a chance of a decent sleep, shower and good meal, and an opportunity to wash clothes and enjoy a bit of personal space away from the general melee of dormitory life. So this is our one night. It is a nice looking place, chalet-esque in character with its wooden roof and red balconies and a pleasant terraced courtyard at the front.

After some negotiations with the owner and a lot of dithering on our part we check ourselves in. It's expensive – about €80 each including an evening meal and breakfast, but it is one of those 'what the hell' moments when one's rational mind is overtaken by the primitive desire for comfort, warmth, food and lashings of red wine. The room is great; comfortable with all the trimmings and there is a balcony with a panorama that stretches back over the Val Ferret, Orsières and southwards across to the Mont Blanc massif – a real room with a view. There is also a rather fine allotment out the back and we can see Madame gathering a selection of vegetables, hopefully for tonight's dinner. As allotments go, you could not do much better than this and it puts my own one back home to shame, neglected as it is. When I first took on my allotment some years ago, Rupert joined me as an 'honorary' associate member and we did sterling work together as we knocked the old, unused plot into shape. However, it was not long before tales of Rupert's earthy efforts became common knowledge and he was seduced by various other friends into helping out on other, more bountiful plots with bribes of cake, tea and as much rhubarb as he could carry. So I was jilted by Rupert, the great journeyman of allotmenteers, who dug, picked, tweaked and nipped his way into other people's cabbage shaped hearts. It's a fickle business, and Rupert – the 'allotment tart' as he is now known within the vegetable community – is to allotments what Ricardo is to the TMB. In fact, I am surprised that he hasn't yet volunteered his services to our good lady outside, her arms brimming with green and golden bundles of leeks, kale and carrots.

Later on we wander down to the restaurant and have a fine meal of beef and fresh vegetables (the very ones we saw being gathered outside earlier) and a couple of carafes of red wine that go down all too easily. The restaurant is deserted, apart from a couple in the far corner, and the radio plays some very traditional sounding Swiss music that even includes sections of yodelling, so our stereotype is complete. It's nice to just sit, chat and eat and helped by the wine we immerse ourselves into the warm, luxurious fug of evening contentment. We talk about books, therapy, marmots, families and thoughts on various future projects. We also consider our plans for the end of the walk in just a few days' time and the various options we have open to us. Rupert's partner, Lucy, is planning to fly down to Geneva and meet us at the end of the walk, or even a little before the end so that she can walk the last day with us, but working out the logistics' of us being in the right place at the right time is quite complicated.

Replete with food and wine, we retire for the night and although it is only about 10.00 p.m. we are both tired. From the balcony we watch a lightning storm brewing in the far distance of the massif; pulsating flashes of bright blue-white light shimmer and twitch, briefly illuminating the white peaks as if a faulty light bulb has been left swinging wildly in the heavens. And then comes the distant sound of thunder that rumbles and rolls and ricochets eerily around the mountains, gently reminding us of their foreboding presence. Somewhere, out there, some serious weather is going on. Somewhere, out there, there will be climbers, mountaineers, farmers and shepherds perhaps, hunkering down in the fragile safety of their little shelters waiting for the storm to pass. I for one am glad that I am tucked up in the cosy warmth of this little hotel.

Nineteen

I slept really well last night, unusually for me, and it was a pleasure to have something of a lie-in and make the most of the comfortable bed, which was very welcome after the rather austere platform beds of the refuges. After an equally luxurious and very long shower, in which I work my way through the hotel's supply of soapy, bubbly, frothy and smelly stuff simply because I can, I join Rupert for a superb breakfast of cereal, yoghurt and nuts followed by bread, cheese and ham. My sore throat and cough has now developed into a full-blown snot-fest of a cold and I have cracked into my emergency supply of Nurofen, normally reserved for dodgy knees and other walking-related injuries. Rupert and I have also not shaved since we left the UK and are both looking rather fuzzy around the facial area. I tell Rupert that I don't generally like beards, but that this one is starting to grow on me. He rolls his eyes and makes a slightly pained expression – a little early for bad jokes perhaps. I do like growing a beard, but don't like actually having a beard. There is a certain critical point in the facial hair process – a tipping point – where enough is enough and it simply has to go. I have decided this morning that beards and a very snotty cold don't go together… say no more. So after breakfast I head into the village to buy a cheap razor and stock up with Nurofen. For some reason I go barefooted, not sure why but perhaps just because it feels good after days of yomping in my heavy walking boots. Stepping out barefooted and with no rucksack feels delightfully liberating, as if I am floating a few centimetres from the ground, and the physical sensation of the stone and grass underfoot is invigorating and I am sure restorative in terms of my circulation.

Barefoot in Champex, I tenderly make my way along the edge of the lake. Despite (or because of) the storm last night, it is a beautiful morning and the sunlight sparkles in the crystal clear water of the lake, around which fishermen maintain a continual presence, motionless like cast iron Antony Gormley statues, gazing into the mid-distance of the lake as the water laps at their wellied feet. As I pass them by I can see keepnets and buckets full of pristine looking fish, trout perhaps. Watching them, I am reminded of the fishermen of Chesil beach in Dorset who seem to forever line the water's edge, night and day, their eyes fixed on the distant horizon. On holiday there once, I recall that they reminded me of soldiers, distant echoes of times gone by as they defend the shores, their fishing rods standing proud where they once held guns. Echoes of war are never that far away.

Here in Champex just past the hotel stand a couple of artillery guns, light cannons surrounded now by wooden troughs of coloured geraniums instead of boxes of ammunition. Unseen behind the guns lays the Fort d'Artillerie de Champex-Lac, an artillery fortress hidden deep in the heart of the mountain. Built between 1940 and 1943 the fort provided the key defensive element of the Great St Bernard area with up to 600 metres of tunnels dug into the mountain, which could house up to 300 men. In later times, during the Cold War era, the fortress was refitted and modernised in preparation for a potential nuclear conflict and the Swiss army maintained a presence there until 1998. It sounds a little like an underground lair from a James Bond film – we just need the top of the mountain to lift off and a huge death ray to slowly emerge and lock onto the sun. Ricardo could be our Bond villain, holding the world's economy to ransom with his great horde of Swiss francs. Apparently, you can take tours of this hidden fortress, maybe another day.

Breakfasted, showered and shaved (well, not Rupert who is still flying the beardy flag) we have a final coffee and consider the day ahead. The main TMB route for the day is a 16-kilometre, 742-metre, five-hour hike to Col de la Forclaz, a route commonly known as the Bovine Trail due to the fact that we pass through Alp Bovine. Of course, more commonly (to us anyway) it is known as the 'cow trail'. However there is an alternative option, the famed Fenêtre d'Arpette route, a much more spectacular and tougher route that our guide Mr Reynolds assures us should only be taken if the weather is looking calm and settled. Whilst a lesser distance of 14 kilometres this route entails a heady climb of 1,200 metres and takes you up to a high point of 2,665 metres, rivalled only by the high point we reached on day two when we crossed the Col des Fours from the Refuge de la Croix du Bonhomme.

By all accounts this is a pretty stunning section of the TMB, a true mountain pass that affords impressive views of the Trient glacier and over the last couple of days there has been much talk among the TMB community about whether or not to do the Fenêtre d'Arpette. The general feeling has been that if the weather is fine, it's a no brainer... what's not to be done about it? Why would anyone pass over the chance to indulge in one of the finest and most spectacular sections of the TMB? The sun is shining, the forecast is good – Houston, we are good to go; everyone is going over the top. Well, being therapists, Rupert and I have decided to take the counter-intuitive approach, the paradoxical intervention one might say. Stuff the Fenêtre d'Arpette, we are taking the cow trail. It looks spectacular enough to us and we also know that it is going to be deserted and the idea of a day's walking in total peace and solitude is simply too good to pass over.

So after a leisurely morning and feeling rested, clean, fed and generally energised we depart the Hôtel Belvédère at about midday, late for us. The route takes us initially through Champex, where we stop to buy a few provisions for lunch, and then out past the La Breya chairlift that does not seem to be operating at the moment. We then head left along a minor road that weaves gently down through a wooded section of pine trees before reaching a more open area of pasture and a cluster of chalets that constitute the small community of Champex d'en Haut. The small metalled road continues along the floor of the valley – the peaks of Le Catogne and Le Génépi rising steeply on either side, and La Breya guarding our rear – and we pass through a further little clustered string of chalets at Champex d'en Bas. This first section takes us an hour or so and is very pleasant as we walk through the aesthetically pleasing valley landscape of pastures, meadows and alpine chalets, the midday sun warming us gently, and we find ourselves sauntering along in a most relaxed fashion. It has been quite a few days since Rupert and I have been alone on the trail together and we enjoy the space and opportunity to simply walk and talk. Ricardo is somewhere in the mountains above us, clambering up to the Fenêtre d'Arpette, but we will catch up with him later at Forclaz, all being well. As we walk, we discuss various aspects of our work: Rupert's family therapy and my play therapy. We hatch plots about setting up a little practice together when we get back to the UK, both of us searching for possible escape routes out of the relentless corporate pressure of our respective organisations. We also talk about various book ideas and all manner of projects and ventures that we could embark on and it is simply a pleasure to be afforded the time and distance away from work to think creatively about this aspect of our lives.

A yellow sign points the way towards Plan de l'Au and Bovine and we leave the road to join a rough 4x4 track that in turn becomes a rugged and very steep footpath that begins the climb out of the valley and up towards Alp Bovine. It's a tough, steep 750-metre climb and the air feels close and humid, even a little tropical at times. This illusion is added to by the thick vegetation, lichen covered rocks and the soporific sound of tinkling water music from the various streams that tumble down from above. All in all it feels a little prehistoric and it reminds me of a childhood film, *The Valley of Gwangi*, about a cursed, lost valley full of menacing dinosaurs. The creatures for this film were created by a great hero of my early years, Ray Harryhausen, the visual effects genius and creator of the stop-motion animation technique, which he called Dynamation, that perhaps came to prominence in his seminal film, *The Golden Voyage of Sinbad*. Some of you out there, of a certain age, will know what I am on about.

Still, I think it is unlikely that Rupert and I are going to be swooped upon by some great leathery pterodactyl, although there once was a very firm belief that the Alps were inhabited by all manner of strange creatures, and especially dragons. Perhaps emerging out of the folklore at the time, this belief in monsters was given credence as late as the eighteenth century by Johann Scheuchzer, a Professor of Physics at Zürich University, who in 1723 wrote a definitive study of the Alps based upon a number of journeys he made around the region. Scheuchzer was clearly a prominent, forward thinker of his day and by all accounts maintained a robust reputation within his field. He also clearly had an enquiring and curious mind, positing some early ideas about the movement of glaciers (the expansion of water trapped in their crevices) as well as indulging an interest in palaeontology where he famously discovered the fossilised remains of a human that had perished in the 'Great Flood', although, in fact, they turned out to be those of a large salamander.

Scheuchzer was a Fellow of the Royal Society of London and corresponded with many scientists of the day, including Isaac Newton, so some may have been surprised when this man of science compiled a compendium of the various species of dragon known to be living in the Swiss Alps. Based upon 'reliable' witness reports, these dragons came in all shapes and sizes. Some were snake-like, some were cat-faced, some had forked tails or forked tongues and probably the most bizarre of these creatures had the head of a ginger tomcat, a snake's tongue, scaly legs, sparkly eyes and a long hairy, two-pronged tail. This is the stuff straight out of Harryhausen's wild Dynamation dreams, and Scheuchzer's book certainly rekindled a fevered imagination very prominent in the Middle Ages that saw dragons everywhere in the Alps. Each region of the Swiss Alps had its own dragon, so the Val Ferret – where we have been walking for the last few days – had a friendly one with a diamond encrusted tail. Hmm, well we didn't see it. The bizarre ginger tomcat dragon lived in the eastern Grisons area of Switzerland, a place that Scheuchzer said 'was so mountainous and so well provided with caves that it would be odd not to find dragons there'. Well, to be fair, most people might think that it would be odder if you were to actually stumble across a dragon in a cave.

Apparently, according to Scheuchzer, one sure-fire way to identify a real dragon from a 'false' one was by the number of birds they had inhaled during flight. Of course, how else would you identify a dragon? Scheuchzer concluded from his studies that 'from the accounts of Swiss dragons and their comparison with those of other lands… it is clear that such animals really do exist'. Perhaps Scheuchzer was being mischievous, perhaps indulging his creative whims or maybe he really did believe in these creatures. These were interesting times, a heady cocktail of theology, superstition and science – the lines between them often blurred and indistinct.

But as we make our way up and out of this valley with its streams of iridescent green water that trip and tumble their way down from the glaciers above along river beds of pale green, moss-covered rocks; the track twisting and turning through lush, verdant vegetation and past all manner of nooks and crannies that have been carved out by years of erosion; and high above our heads the constant presence of the peaks that seem to disappear into the heavens, it is easy to understand that back in the Middle Ages and before the Alps must have seemed fearful and unknown. So it is unsurprising then that tales of dragons, witches and dark magic would abound; that myths and legends would come to define a place that for so long was dark, inaccessible and dangerous. Indeed, it seemed that every valley and probably every village had their own particular stories, legends and superstitions; passed down through the generations and told around dark, smoky hearths as the weather closed in during the long winter months. Tales of giants, devils, fairies and ghosts were numerous as were those of little creatures; magical dwarves, gnomes and imps that were said to swarm in abundance around the mountains, hiding away in crevices and caves.

The people of Chamonix in times gone by used to sometimes refer to Mont Blanc as Mont Maudit, the 'accursed mountain' and in 1690 the villagers even paid the Bishop of Annecy to exorcise the glaciers that threatened to destroy their houses. Apparently it worked as the glaciers retreated by about 200 metres, with the bishop subsequently hitting the villagers with a massive bill for his services. (Crikey, just think what he could charge now with an impressive little bit of global warming assisted exorcism.)

Northeast from here along the Trient valley is a little village called Les Diablerets – the 'abode of the devils' – that suggests many a dark story. One of these stories from Les Diablerets is about the Tsanfleuron glacier, which translates as 'field of flowers' in the local dialect because, according to legend, the glacial ice sheet was once a lovely sunny meadow. One day some mischievous devils arrived in the mountains to play skittles with great boulders that they threw around with reckless abandon, some of which would accidently strike the nervous shepherds who had bravely ventured up to tend their flocks in the high pastures. Eventually, the shepherds had enough of dodging the great rolling boulders and abandoned the pastures so that gradually, because of the playful devils, the Tsanfleuron was forever transformed from the field of flowers it once was into a boulder-strewn icy wasteland, as it is today. Other stories from Les Diablerets talk of the sorrowful groaning of lost souls that foretold terrible landslides in 1714 and 1740 and of a type of imp who one night turned himself into to a fox and sat up all night in a hayloft knitting with the hair of his own tail. As you do.

One of the most compelling and enduring of alpine legends concerns that of Mount Pilatus, which towers over the waters of Lake Lucerne in central Switzerland. Word has it that this mountain was named after Pontius Pilate, who according to legend rises on every Good Friday from a lake just below the peak, sitting upon his great judicial throne and dressed in the fearsome blood-red robes of office. Anyone unfortunate enough to lay eyes upon this ghostly apparition is fated to die within a year of the vision. The stories behind this legend are many and various, as they ever are, but the prevalent myth has it that Pontius Pilate, fearful of facing punishment or even death at the hands of Emperor Tiberius, committed suicide in Rome and his body, weighed down with heavy stones, was thrown into the river Tiber. This act was followed by a sudden spell of the severest storms Rome had ever seen and so, associating this with the cursed watery presence of Pilate, his body was summarily recovered and taken to Vienne in south-eastern France where he was hurled into the Rhône. However, once again further terrible storms ensued so he was fished out and dumped into the waters at Lucerne, but here too the malevolent spirit of Pilate whipped up a ferocious storm so they dredged the poor man up a third time and tossed him into a lake high up on the mountain that towered over Lucerne, far enough away to lie undisturbed and not cause too much more trouble.

Life returned to an air of normality and relative calm after this, although every year a number of people would mysteriously die from unknown causes. These deaths were attributed to the fact that the victims must have set eyes upon the blood-red ghoul rising from the lake. And every now and then terrible storms and blizzards would rage around the mountain, caused no doubt by people taunting Pilate by throwing stones into the lake.

So the peak became known as Mount Pilatus (although it is far more likely that it actually took its name from the Latin *pileatus,* which means capped – by clouds in this case) and for many centuries the authorities forbade anyone to approach or climb the mountain for fear of disturbing the troublesome spirit; several people even being imprisoned for doing so. It was almost two hundred years later in 1555 that the naturalist Conrad Gesner climbed Pilatus and celebrated his achievement with a spirit defying blast on his alpenhorn, and in 1585 the pastor of Lucerne, Johann Müller, in the company of a plucky band of townspeople, further challenged the evil ghost by climbing the mountain, throwing stones into the supposed cursed waters of the lake and even wading in to provocatively churn up the waters. Brave people, but there was no response from the 'other side' so it seemed that the centuries old spell of Pontius Pilate was finally broken. But just to be sure, in 1594 a gap in the wall of the lake was dug out by the people of Lucerne, draining the water and it was not until 1980, some 400 years later, that the gap was eventually closed and the lake returned to its former state. So, who knows? Climb this mountain at your peril, that's all I can say. And it goes without saying that Mount Pilatus also has its own dragon. History records several sightings of this beast, most notably on 26 May 1499 when, following a terrible thunderstorm, several 'well respected and educated townspeople' witnessed the spectacle of an enormous, wingless dragon rising out of the river Reusse, its water having been washed down from Pilatus by the great storm. Hmm, I don't know about you but I am starting to wonder about those aforementioned mycological societies. Perhaps the locals have been celebrating the wild mushroom a little too much in these parts.

And talking of Mount Pilatus, mention should also be made of the fact that the mountain is home to the steepest cogwheel railway in the world, with a maximum and somewhat daunting gradient of 48 per cent, averaging at 35 per cent. To put this into context the gradient, at its peak, is steeper than any existing street in the world with the train climbing a slope of more than 1,600 metres in a mere 4.6 kilometres. And for those of you who might be interested, the record for the steepest street in the world is claimed by Baldwin Street in Dunedin, New Zealand, which comes in at a pretty stiff 35 per cent – not a place that you would necessarily choose to live. Every summer, the street plays host to the annual Baldwin Street Gutbuster, a race that involves running to the top and back down again in the fastest possible time (the record currently standing at one minute 56 seconds) and as something of an occasional hill runner myself I quite fancy the idea of the challenge. When the Pilatus line first came into operation in 1889, using steam traction, it took over an hour to reach the summit, averaging a speed of around four kilometres an hour as it clawed its way up the side of the mountain. Today, the electrified cars run at nine kilometres an hour and reach the summit in close to half an hour, precariously hugging the side of the mountain. Devised by Eduard Locher, it was an impressive and unique feat of engineering and the line still uses the original rack rails that are now over 100 years old. There was some concern that the rails had understandably worn down over this period but, a little like an old mattress, it was discovered that this problem could be simply rectified by flipping the rails over, providing a surface that would be fine for another 100 years. Probably about the same timescale that most of us employ for mattress rotation.

But railways and mattresses aside, I love these old stories; folk-tales, legends, myths and superstition passed down from generation to generation, parent to child. These are the stories that bind and weave communities together creating a sense of place and time, adding a rich cultural history and narrative depth to these small villages that for so many hundreds of years have dotted the valleys of this challenging, alpine landscape. These are stories that helped people to make sense of their world. These tiny and often isolated places, without the fundamentals of transport, power and communication that we now take so much for granted, were at the mercy of the often savage conditions; an elemental orchestra conducted by the peaks of the Mont Blanc massif that could move between moments of sublime serenity and wild cacophony in an instant. It is little wonder then that these folks, huddled together around their fireplace in the long winter months listening to the sound of cracking ice, groaning glaciers, the roar of avalanches and the vicious winds whistling down through the valleys from the high peaks, told stories to each other to entertain, understand, explain and engage with the world around them. As a child therapist, I have always been fascinated by the stories that we tell – as children and as adults. There is a timeless quality to storytelling; it is part of the very fabric of what it means to be human, how we understand our relationships, the world and ourselves. It is about narrative identity; by what means we express aspects of our experiences and ourselves, and as Rupert and I make our way on this journey around Mont Blanc there is a sense of us just gently brushing against the rich, timeless narrative history of this mountain landscape.

Eventually, we emerge out of the valley and through the treeline, and the stony, well trodden path curves gently across a broad slope of open, gently contoured grassland that is sprinkled liberally with yellow and white flowers, a late autumnal flourish of floral colour before the winter snows arrive, which will be all too soon in these parts. The view to our right drops away dramatically with the town of Martigny far below to our northeast, miniaturised by the distance, and beyond that the great flattened plain of the Rhône valley that sweeps away into the beyond, the textured patchwork of greens and browns an indication of the intensive agriculture along this fertile valley floor, a far cry from the tiny farming communities up here in the heights. There is an almost unworldly quality to the view of the Rhône valley from up here, as if it were some immense landing strip cut through the mountains and I am reminded of Erich Von Daniken's book, *Chariots of the Gods*, which I read as an impressionable adolescent all those years ago. In Von Daniken's fairy-tale world, this great swathe of valley would have been prepared by some ancient civilisation in readiness for a visitation by godly astronauts from another world, cryptic symbols scratched into the baked earth. It's a stunning view and so in tribute Rupert and I decide to take a break and stop for a bit of lunch, having come across a lovely spot by an old tree that is calling out for a bit of company. Back in Champex we bought bread and cheese and a couple of tomatoes and once again the simplest of meals tastes like a feast, earned through four hours of hard slog. This humble bit of food, up here on the slopes of Alp Bovine, with the view to the snowy peaks of the Grand Combin to the southeast and Mont Blanc away to the southwest, tastes better than any three course meal in some fancy restaurant back home. Context is everything, as they say.

After lunch we press on and it is only another 20 minutes before we reach Alp Bovine itself, which sits at 1,987 metres on the pastured slopes facing down towards Von Daniken's Rhône valley runway. There is a little dairy farm here and we have to make our way past some seriously mean looking cows, great black muscled beasts with thick white horns tilted intimidatingly towards us. Around their necks, which are almost as thick as their bodies, a studded leather collar holds an enormous cow bell that gives out an occasional deeply toned clunking chime as the cow gives a casual flick of its enormous head, weary of the flies buzzing around its fulsome face. These are Hérens cattle, a particular local breed named after the Val d'Hérens region of the Swiss Valais. They look like they are wearing toupees, but despite this comedic touch they are stocky, robust, proud looking creatures and apparently well known for their feisty female nature, part of which involves a fair degree of tussling as they seek to establish a hierarchy within their herd. I feel a little nervous of them as I tread gently by and hope that they don't see me as too much of a rival arrival, so to speak, but Rupert is much more casual as he saunters through the herd, chatting to them amicably as if he were Dr Doolittle. It wouldn't surprise me if he stops for a while and asks one of them to pose for a quick sketch. As we pass by, one of the cows adopts a rather defensive posture and I ask Rupert how fast he can run. Not fast enough he suggests and I say something about it being a shame that his rucksack is so heavy.

As a result of the Hérens' tendency to lock horns with each other, a rather unusual tradition of cow fighting has grown up in this region of Switzerland. The first of these bovine battles took place on the Col de Balme above the Chamonix valley in 1917 when the Hauts Savoyards challenged their Swiss neighbours, and today a handful of breeders maintain this annual tradition, known locally as the Combat de Reines – the Battle of Queens. The most recent battle took place last year in La Tour, just a stone's throw from where the TMB will be taking us tomorrow, when close to 2,000 people turned out to watch a young Hérens heifer be garlanded with flowers and crowned Queen of the Alps.

I am not sure where I stand on Swiss cow fighting – and that's not a sentence I ever thought I would write. The cows are clearly unharmed in the process and perhaps they are only being corralled into doing what they do naturally in the herd as they maintain their group hierarchy, akin to how horses are just doing what comes naturally in the 3.30 at Doncaster. The tradition seems to have become part of the regional practice of cattle trading, a way of displaying the cattle with the winners being sold for higher prices at auction. Certainly, they are fine looking animals and well looked after but, truth be told, I am not really a fan of any animal performing for human entertainment, whatever form it may take. Anyway, these Hérens are thankfully docile enough as we make it through the herd unharmed, and just a little further on we see the farm building and a small refuge that offers beds if needed and a cafe/bar that provides drinks and food for locals and hikers. It is a surprise to see this little stop-off up here and had we not just stopped for our own lunch we might have been tempted to avail ourselves of its hospitality for a short while.

A short distance from the farm buildings, the path leads up to a wooden gate in a fence that borders a wooded section of larch and pine. This is the Collet Portalo, our highest point of altitude for the day, that sits at 2,040 metres and before going through the gate we take a last lingering glance back towards the rounded meadows of Alp Bovine and the shimmering peak of the Grand Combin that floats like a distant isle atop the blue-white stippled sky beyond. From here the trail descends pleasantly among the forest, the path gently meandering through the trees. Mostly even but in places rocky and root-covered, the path is agreeable and a welcome respite from the long, hard climb up from Champex. Since setting out this morning we have met only one or two other hikers on the trail and the experience of walking in relative solitude has been an absolute delight. Rupert and I chat at times, but are also comfortable enough in each other's company to walk in silence, simply enjoying the quiet stillness of the countryside, interspersed with the occasional friendly chatter of birdsong from the trees and the latent summer buzz of insects from the bracken hedgerow. Every now and again, tempted by the macro-mode of my camera, I stop to take photos of large black beetles with spindly antennae, outsized grasshoppers and speckled, ochre butterflies resting gently upon the delicate blue petals of mountain gentians. The warming sun still maintains its daily vigil in a clear sky, as it has done since we first set foot in Les Houches and while I am sure the variant route taken by Ricardo *et al.* via the Fenêtre d'Arpette has been stunning it would be hard to better our day's walking as we make the gentle approach to Col de la Forclaz. At one point, we cross a magical glade dotted with cairns; stone markers piled a metre high and we wonder what they might signify. They are clearly not route markers and we wonder if this might be some kind of ancient mountain cemetery from days gone by.

After a while the path emerges out of the woodland and follows the edge of a number of large, fertile pastures with more of the black Hérens cattle that follow our progress with what looks like disdain, although to be fair it is never easy to read the expression on a cow's face, inscrutable creatures that they are; they might be excited, curious or mildly aroused for all we know. They would make good poker players that's for sure. Forget cow fighting – cow poker! Now that would be something I would pay to watch, despite my best reservations. A short distance from Col de la Forclaz we meet a couple of fellow hikers, one of whom has clearly had a nasty fall and badly gashed her leg which has been hastily tended to with assorted bloodied bandages. The man is carrying both their rucksacks allowing the poor woman to gingerly make her way along the path. We stop and offer assistance, but they insist they are fine and only have a short distance to go, but it is a reminder of how easy it is to slip and injure oneself on the often uneven, rocky track.

Before long we see the Hotel du Col de la Forclaz up ahead, a pair of large, imposing buildings that house 38 beds in four separate dormitories, as well as a restaurant and cafe. It has been a long day on the trail and Champex feels a million miles away, rather than the rather paltry 16 kilometres that we have actually done. Wearily slipping off our rucksacks and pulling off our hot, sweaty boots, we negotiate the process of checking in then head to our dormitory only to find the mighty Ricardo already in residence. 'So how was the cow trail?' he asks, with a playfully wry grin. He feels worthy, as he should, having taken the tougher route up the Fenêtre d'Arpette, although we do point out that his route was a mere 14 kilometres. But having stayed in the comfort of the hotel back in Champex last night and opting for the less challenging cow trail today, Ricardo has us down as casual part-timers rather than hard-core hikers like himself, perhaps Bourrit to his Balmat. He enjoys teasing us, but it is with good humour and it is great to hook up with him again – a good friend on the trail.

After marking out our territory in the bunkhouse we retire for a couple of beers on the terrace and Ricardo fills us in on his trip over the top, which sounds eventful and dramatic. From his account, the *variante* route up to the Fenêtre d'Arpette was tough going, with a climb of 1,200 metres and a descent of 1,140 metres but it sounds like it was worth the effort for the stunning views of the Trient glacier. Ever the consummate networker, he has hooked up with a number of other hikers and so our Chaucerian band of merry travellers continues to grow and there is a pleasant hubbub of TMB chatter around the tables as the beers slip down all too easily. We still miss the lively presence of Julie and Claire, but reports are that they are doing well, Claire has managed to meet up with her boyfriend somewhere along the route and plans are still afoot for us all to meet up in Chamonix, just two or three days away now.

The hotel itself, more of an upmarket refuge, sits in the Col de la Forclaz pass between Martigny and the French border, the terrace is next to the main road that sweeps through on its way to Chamonix. History has it that the pass was originally a mule track used by both travellers and smugglers as they weaved their way up the mountain from Martigny and across the border into France. Little known in the nineteenth century, the area didn't become popularised until the rise of tourism in the region, especially in nearby Chamonix, and now on the road that wasn't opened for use by private vehicles until as late as 1920, the mules are left for dust as shiny chromed bikers speed by in noisy clusters, leaning dramatically as they take the sharp corner by the hotel. It must be a dream road to ride and with its relentless hairpins and sharp ups and downs it apparently rates as one of the best (and scariest) on the circuit. I don't envy these bikers as they accelerate away towards Chamonix. What will take us two days may take them a mere 30 minutes, but give me two feet, a stick, a silly hat and a mountain to climb any day. I prefer to acquaint myself with the land upon which I am travelling; to feel my feet upon the warm earth and experience each step on the rough, well worn path.

It's a bustling place, clearly popular, no doubt a result of the bar and restaurant that draw in a much wider community than just the TMB crew. The hotel was apparently first built around 1830 and has been in the same family for six generations. Some notable characters have stayed here over the years, namely Johann Wolfgang von Goethe, Edward Whymper, Victor Hugo and Alexandre Dumas among others so it has quite a cultural pedigree. Later, in the restaurant, we have a wonderful meal of vegetable soup followed by chicken, chips and cauliflower. When we finish, the waitresses come round with an offer of seconds and we duly load up our plates and start all over again and the carbohydrate/protein overload is just what we need after the day's exertions on the trail. My cold seems to have worsened and although exhausted I dread going to bed, both because I know I feel worse at night but also because I am acutely aware that a snotty and unpleasantly noisy cold is not especially welcome in a dormitory situation. Thankfully the place is not too crowded and I manage to have a spare bed on either side of me (perhaps no coincidence) so I use the space to its optimum, sprawling out and drifting into a fitful sleep.

Twenty

I walk therefore I think; I think therefore I am. Therefore I walk and… I am. This, in shorthand, is my philosophy about these long distance hikes, the sense that in a funny kind of way I walk myself into existence. Robert Macfarlane, in his wonderful book *The Old Ways: A Journey on Foot*, talked about his sense of '… walking as enabling sight and thought rather than encouraging retreat and escape; paths that offer not only means of traversing space, but also ways of feeling, being and knowing'.[17] The well worn path that Rupert and I and so many others are following around this mountain has literally been walked into the ground over many hundreds of years – it connects us together in body and mind, to the past, present and future. The path leads us around, but also back. It takes us both outside and within. We are all time travellers, metaphysically, as we move back and forth along a continuum of past experiences and possible futures, just as walking leads one to all sorts of junctures, crossroads, encounters, signposts and route markers. Paths, as Macfarlane says, run through people as surely as they run through places: 'walking is not the action by which one arrives at knowledge; it is itself the means of knowing'. Under the ever watchful eye of Mont Blanc – La Dame Blanche, Il Bianco, call her what you will – who has come to personify so much, an infinite narrative drama of human possibility is enacted by the many thousands who make this journey – each in contemplation to some degree about the personal path that they are walking into existence. And I am acutely aware now that the end of this particular journey is not far off, perhaps two or three days, and that this little joyful bubble of reflective space is soon to be punctured by the reality of everyday life, a harsh intrusion in some ways, but not so in others. It is as if real life, whatever that means, has been temporarily put on hold.

This morning, daylight is a harsh intrusion and I wake up feeling particularly rough, full of cold and deprived of sleep. Nevertheless, I am slowly revived by a hearty breakfast of croissant, toast, jam and Nurofen, all washed down with several cups of strong coffee. Today we have what looks like a challenging six or so hour stretch to Tré-le-Champ, across the Col de Balme and back into France. It is only 13 kilometres overall, but includes an overall height gain of 1,069 metres and a descent of 1,178 metres, so it is going to be a long day one way or another. We head off at around 8.30 a.m., the path picking up from outside the front of the hotel following the Bisse du Trient and signposted down towards Le Peuty. The *bisse* in question is in fact one of many man-made channels created throughout Canton Valais with the purpose of carrying glacial meltwater over often considerable distances to irrigate agricultural land, including the fertile plains of the Rhône valley that we were contemplating yesterday. These water channels, laboriously cut out of the bare rock, date back many hundreds of years with the oldest going as far back as the twelfth century, and in places where the geology of the terrain meant it was impossible to excavate, hollowed tree trunks were used to bridge the gaps. Over the course of time, footpaths were created alongside each *bisse* as a way of enabling ongoing maintenance, but in fact the Bisse du Trient once had rails laid along the track to allow blocks of ice to be transported down to the Hotel du Col de la Forclaz from the glacier. Ricardo tells us that they followed the path of the *bisse* for the last couple of kilometres of their descent yesterday and explains that in certain sections the water seems to be running uphill, perhaps an optical illusion caused by the lay of the land.

The path drops down through a wooded slope until we rejoin the road briefly before passing through the little hamlet of Le Peuty and the rather austere, functional looking Refuge du Peuty. Not having had the chance to get anything for lunch we attempt to blag some bread from the refuge but to no avail and the little shop in the village is closed, so we press on with our meagre provisions of fruit, nuts and a few biscuits purloined from the hotel. As we head out on the track that follows the floor of the valley before turning to cross the shallow trickle-water of the Nant Noir we hook up with an Australian family that Ricardo walked with yesterday and who have now been assimilated, Borg-like, into his TMB cartel. They are indeed a lovely family, friendly and easy to be with and we are impressed by the fact that they have somehow persuaded their two teenage daughters to come on this challenging hike. Another family – a charming South African couple with their two children, who must be about nine and twelve – who we have walked with occasionally since Courmayeur, also join us. During our brief encounters together over the last few days we have paused to chat and joke for a while and the father has a great sense of humour as he coaxes, cajoles and drags his family up mountain after mountain. On the quiet he tells us that it is his wife's birthday tomorrow and he is trying to sort out a surprise birthday cake at the next refuge – a romantic and logistic masterstroke of negotiation if he manages to pull it off. And so for this stretch, our group has grown in number, bound together by camaraderie and a sense of collective endeavour.

After a short while we enter woodland and the path starts the inevitable zigzagging up the side of the valley and our group starts to spread out and disperse as people find their natural pace for the long climb up to Col de Balme. Despite my cold I feel physically on great form and start to pick up pace as I hit the beginning of the ascent through the woods. Somehow, I relish these climbs – much more so than the downhill sections of the trail – and after eight days of walking I finally seem to have found those elusive 'mountain legs' and more than that, my body, mind and spirit suddenly seem to fall into step with one other and I feel perfectly synchronised. Leaving Ricardo and Rupert behind I begin to power up the path that weaves its way through the wooded flank of the valley, a perfectly automated machine, my legs like hydraulic pistons as they propel me ever upward. This is the 'flow' moment that I spoke of before and blimey it feels good. This is my mountain and I am determined to take it in one go without pause or respite. I pass several people on the way, offering a fleeting greeting but not stopping to chat – fearful that if I stop I may lose my momentum.

I am falling up towards Col de Balme, and in the intensity of my focus the colours and sounds of the mountain do indeed blend into one, a visual and aural alpine infusion of the senses. For the first time since we set off from Les Houches there seems to be a change in the weather as greying clouds of rippling cumulous begin to fill the blue sky above the mountain peaks and a sharp morning breeze riffles the forest covering of pine, spruce and larch, which offer up a periodical sweeping sigh in response, nature's wind chimes. On the ground the occasional grasshopper, startled by my footfall, launches itself to safety with a gracefully urgent flickety buzz of leg and wing, often with a flash of red or blue, and the occasional chatter of the birds ricochets among the trees as they communicate my rude interruption.

The path is well worn and easy enough to follow, but still requires a degree of concentration as it is all too easy to trip or stumble on the rocks or knots of twisted roots that invariably mark its progress. Eventually, I emerge from the treeline and every now and again, as the path switches back on itself, I can see the distant figures of Rupert and Ricardo below me and give them an encouraging wave. Out of the trees, the terrain once again opens up into a broad, expansive landscape of sloping green-brown alpine pastures, a little like Dartmoor or the Lake District apart from the ever present layered, theatrical backdrop of the snowy, ice-bound peaks that rise up into the sky.

Before long the relentless gradient begins to ease up and the path meanders its ascent to the Col de Balme, a great sweeping saddle of a mountain pass that marks the border between Switzerland and France and lies almost equidistant between the two attendant peaks of Tête de Balme and Les Grandes Otanes. This is the spot where the practice of cow fighting began, but thankfully there's no sign of any bovine wrestling today. This is also pretty much the furthest point on the trail from the summit of Mont Blanc, which lies far to our southwest, and having been hidden from sight for much of the last few days it is good to see our old majestic friend once again. The pass is dominated by the solid presence of the privately owned Refuge du Col de Balme, its austere grey exterior broken up by the vivid red window shutters that today are firmly closed, quashing our hopes of a coffee or even early lunch. There has been a refuge here since 1877 and as a consequence of straddling the border between France and Switzerland it has apparently been caught up in many a cross-border dispute and even been burnt down and rebuilt several times. For the time being, the refuge sits a metre into Switzerland but by the sound of things that could well change at some point in the future, depending which way the proverbial borderline wind might blow. As it is, up here at 2,191 metres the actual wind is blowing pretty strongly, bringing with it quite a chill factor and so along with a sizable group of other walkers who have congregated here at the pass, I seek shelter by propping myself against the wall of the refuge and catching my breath after the exertions of my non-stop dash from the valley below. After ten minutes or so Rupert and Ricardo arrive along with the various other members of our 'group' and we take a little time to rest, recuperate and plan the remainder of the day's route.

According to Mr Reynolds, our trusty guide, the main TMB route to Tré-le-Champ takes us via the summit of L'Aiguillette des Posettes and should not be taken if there is the possibility of a storm brewing, presumably because of the danger of lightning strikes. I recall once getting caught out in a lightning storm with my friend Rob when walking a high ridge section of the GR10 and we literally had to run the last hour of the stage to keep ahead of the storm and get to the safety of our refuge before we were toasted. 'Never take risks with mountains,' Rob would tell me; this after I had jettisoned my survival bag, thermal blanket and any other safety related equipment in a bid to keep the weight of my rucksack down. Rightfully, he despaired of my lackadaisical approach to mountain walking and personal safety.

Up here on the Col de Balme we make a brief assessment of the weather conditions and come to the conclusion that it is fine to push ahead on the main trail. The weather does seem to be turning, but there is still plenty of blue sky about and it will only be another couple of hours before we are away from the high ground and heading back down into the valley towards the safety of Tré-le-Champ. So having negotiated the multitude of route markers and finally getting our bearings we step across into France and drop down through sweeping grasslands, hugging the contours of the Tête de Balme before climbing back up towards the broad saddle of Col des Posettes. Again we are confronted with a yellow route marker signpost with a bewildering array of possible options as various paths criss-cross each other, but we pick up the trail and head downwards once again before making the final ascent of the day up to L'Aiguillette des Posettes.

Ricardo is on fine form and still practising the otter joke that I told him three days ago, although a part of me is wishing that I had never mentioned it. But to be fair, he is applying himself diligently to the task and looks forward to impressing Julie and Claire with his newfound comedic skills. He is a little worried that being younger than us, the girls may not have heard of *Tarka the Otter* and that after all his practice the joke will be rendered meaningless, but we assure him that with their RSPB associations and overall hard-core environmental credentials, Tarka will have been an indelible part of their childhood. We hope anyway. As we walk, we pass what looks like some kind of ski station and then rising up ahead of us the craggy crest of L'Aiguillette des Posettes, our high altitude point of the day that sits at 2,201metres. The trail follows the ridge and sometimes breaks up into several smaller paths that twist and turn their way through the grass and heather, dotted with rocky outcrops, but the route up to the summit is clear and we pick our way through. As we make our approach towards the peak we walk through great swathes of alpine blueberry shrubs, the leaves of which have turned a deep autumnal orangey-red, dramatic in their vividness and startling in their contrast to the greys, greens and whites that tend to dominate the scene. The shrubs carpet the ground in a wash of vibrant colour, truly spectacular, and lend a visually dramatic quality to the scene, only added to by the stunning 360-degree panorama that opens up as we reach the summit.

The view from L'Aiguillette des Posettes, the summit marked by a tumbling cairn of rock, simply blows us away and this has to be one of the best and most dramatic scenic stages of the walk so far. We are standing pretty much at the northern end of the Chamonix valley, almost but not quite full circle, and we attempt to take in the entirety of our surroundings. To the north, at the head of the valley, we can look down towards the small settlements of Vollorcine and Les Rupes (which I tell Rupert should really be his home town) and then sweeping south along the valley lie Argentière and Chamonix, which is our final destination. In the foreground, the peaks of Aiguille Verte, Aiguille du Chardonnet and Aiguille du Tour all compete among others for attention while the snow-encrusted dome of Mont Blanc sits regally to our southwest. Several of the peaks are topped with strange flying saucer shaped clouds and to our left the greying wall of the Glacier du Tour hugs the rock in its perma-clasp icy grip. Ahead of us we can just see the flat reservoir waters of the Lac d'Emosson that hangs incongruously high in its amphitheatre of rock, the 227-million-cubic-metres of water held back by a huge white walled dam that from this distance almost looks glacial itself, wedged incongruently into the rock of the mountainside. It feels like we really are on top of the world and we are reluctant to move on, knowing that the descent back down into the valley and towards Tré-le-Champ is really the beginning of the end of this journey; that it is unlikely that we will have an experience to better this. And so we hang around for a while, with half an eye on the gathering clouds as we just soak up the peaceful ambience of the place, each of us lost in our own thoughts. Other walkers drift by, hang around for a bit and then move on and eventually we also decide to head off, aware that if the weather does change we don't want to be caught out up here.

The route from here follows the ridge-line along a rocky track and then starts to descend quite steeply, helped or perhaps hindered by a series of sections of wooden steps that actually require more cautious navigation than the usual path. Before long the route turns away from the ridge and zigzags down the flank of the hillside through more of the familiar alpenrose, juniper and bilberry and eventually back into the tree covering of fir and larch, through which the path twists its tiring way; relentless and unforgiving upon our weary feet that are just beginning to allow themselves the anticipation of some rest at the end of a long day. As ever, these final stages of the day always seem the longest and the descent seems to go on forever until we finally drop down onto a main road, which we follow for a short while before cutting across through a series of tracks and alleys into the little hamlet of Tré-le-Champ. Having called earlier, we established that the gîte here in the village, the Auberge la Boerne was full and so have booked ourselves into another gîte called Le Moulin, which is in the next small hamlet of Frasserands, just ten minutes down the road. We seem to have lost touch along the way with our Australian friends, whom Ricardo thinks are staying down in Argentière. But for the last couple of kilometres we have hooked up with a young American couple and so we saunter along together, grateful to be walking on a flat road surface, until we finally reach the gîte, an inviting, pretty looking building with a nice terraced garden from which we are greeted with a cheer by our South African friends who have beaten us to it.

Le Moulin is comfortable and less austere than some of the more functional refuges that we have stayed in and the place is run by a cheerful chap who is busy in the kitchen cooking tonight's dinner. As the name suggests, Le Moulin is an old mill dating back to the eighteenth century and there is a worn, solid and rustic charm about the place – all wood and stone. Being smaller than most it sleeps up to 38 people, but rather than big dormitories there are just six bedrooms, each with four to ten beds. All in all it is more personable and homely and we are shown up to a small room with six bunks, where two Dutch girls are also in residence, and after settling ourselves in, showering and washing a few clothes Rupert and I go down to the garden for a much needed beer. Meanwhile, Ricardo has set off for the little town of Argentière, about 15 minutes walk down the road, to finally offload his wedge of Swiss francs. Over a couple of beers, we meet a gentle and unassuming chap called Steve from London who has been walking the TMB on his own and we also have a good chat with the South African dad, whose name we have not quite established. It turns out that back home he is a lecturer in chemical engineering and we swap notes on the relative ups and downs of life as a university lecturer. Whispering conspiratorially, he also tells us that he has managed to engineer the owner of the gîte into baking a cake for his wife's birthday and his obvious delight in this feat of negotiation is clear to see.

As we sit in the garden chatting, reading and writing our journals, the weather finally begins its threatened about-turn and a few drops of light rain start to fall, the first we have had since we set out from Les Houches nine days ago. Can it only be nine days? It feels like we have been walking for weeks. Time is both constant and relative I suppose, and it can pass both quickly and slowly. The theory of relativity suggests that the measurement of various quantities is relative to the velocity of the observer, so does this mean that space and time can stretch, dilate or be pulled out of shape? All I know is that Rupert and I have been travelling very slowly, our passage through space and time measured by the pace of our feet upon the trail and the sun in the sky. We have rarely been aware of the actual time as we have walked and have often lost track of which day it is. I relish this sense of losing ourselves in a dreamy state of relative subjectivity, it's one of the pleasures of long distance walking, but as we close the loop on Mont Blanc we are both aware that we are going to have to make the transition back into a different reality; that once again the demands of our professional lives will come to the fore, hitting fast forward and crunching us reluctantly through the gears until before we know it the weeks will once more be hurtling by. Life is short, something of which I am tangibly aware as I approach 50, and we need to slow it down when we can.

Later, we are gathered together for dinner and our friendly gîte man has knocked up a superb meal beginning with home-made bread and salad and followed by a delicious cheese and spinach lasagna cooked in large stoneware dishes that he casually distributes among the tables. It is probably the best meal we have had on this trip, and this guy who almost superhumanly manages to run this gîte single-handedly amazes us. He brings round second helpings, eager for everything to be eaten and then for dessert we are treated to a delicious home-made cake. It seems that rather than just bake a cake for our friend's wife's birthday, he decided he may as well bake a cake for everyone, so we all tuck in and a rousing chorus of happy birthday lifts the roof off the gîte.

After dinner we sit and chat for a while in the lounge and then, tired and sated, head for bed. As I lay in my bunk, trying to sleep, the rain begins to fall outside, heavier now like a comfort blanket of white noise followed shortly by the rumble of thunder that ebbs and flows, Doppler affected, around the mountains like a game of sonic pinball played off the rising walls of the Chamonix valley. Securely cocooned in the safety of our wooden bunk beds, the storm wraps its electrical charge around us, the window of our room spasmodically brightening with neon flashings of lightening white and I find myself illuminated with mental images of the surrounding peaks: the Tour, Chardonnet and Points des Grandes that tower over us in the brooding darkness of the storm. Somehow, this morning feels light years away, time stretched, and I replay parts of the day's journey from Col de la Forclaz as like little ants we make our slow progress from one valley to another, insignificant within the magnificence of the mountains. 'Puny humans' as the Incredible Hulk would say.

Twenty-one

We awake in the morning to the relentless sound of rain pounding the roof of the gîte and a glimpse through the window reveals a blanket of grey stretching as far as we can see. Well, we have had it good so far and can't complain too much I guess. Downstairs, I join Rupert, Ricardo and London Steve for a hearty breakfast of bread, croissants and jam and the mandatory coffee, the standard morning fuel injection to kick-start us into action. A quick glance outside the front door confirms the steady fall of rain and the temperature seems to have dropped considerably, with reports of snow as low as 1,500 metres. In just a matter of hours we have moved from what felt like a balmy late summer to the depths of winter. Over breakfast we consider our options. This penultimate stage of the TMB takes us up to the Grand Balcon Sud, perhaps one of the finest and most scenic sections of the trail that runs along the north side of the Vallée de l'Arve. Our kindly pocket guide Kev Reynolds says that we should 'pray that the weather will be kind enough to allow us to enjoy these stunning views'. Well, I think we can safely put our bibles away and save our prayers for another day, but then we should be thankful for the wonderful weather that we have had over the last nine days, which has been better than we could have ever hoped for. This section also heralds the infamous *passage délicat* – the delicate passage – a stretch that involves a series of near vertical metal ladders, handrails and platforms bolted into the rock, best to be avoided by anyone who may not have a great head for heights. Even before setting out on this trek I have been looking forward to this section, being slightly drawn as I am to the idea of risk, but I hadn't quite envisaged doing it in sheets of freezing rain. As it stands, we have three options.

Option one: we could go down the road and catch the bus into Chamonix, find a nice dry bar, have a few beers and gaze dreamily up at the Grand Balcon Sud and say to ourselves, 'You know, anyone would be nuts walking up there in this weather.' Rupert is sorely tempted by this idea and I have to remind him of the Faustian pact we made back on New Year's Day in Cornwall as we walked the muddy coastal path to Falmouth: that we are in this together, for better or worse. 'You're the man who set a dog on fire,' I tell him. 'You don't do things by halves. Come on, do it for Toby.' It's a low blow I know, but with being this close to our goal the gloves are off. 'And anyway,' I say (and this is the real low blow), 'what would Ricardo say?'

Option two: take a TMB *variante* that avoids the ladder section and takes a lower route along the side of the valley. Safe, but boring; we would never live with ourselves.

Option 3 is my preferred course of action and is a case of 'What the hell, let's just go over the bloody top and see what happens.' Somehow I will feel terribly cheated if I miss out on this section and know I will come to regret it, but Rupert (despite paroxysms of Toby-induced guilt) is seriously wavering and we begin to make half plans to part company and meet tomorrow in Chamonix where he is also due to meet his partner Lucy who has gallantly flown over from the UK to meet us. However, in the midst of our negotiations a fourth option suddenly emerges that may be something of a compromise. This stage of the main TMB route takes us to the Refuge la Flégère, from where a cable car just happens to run down to the small village of Les Praz de Chamonix, which means that we can walk this section, hop on the cable car this afternoon and both get to Chamonix either tonight or tomorrow. We started this journey together at 5.30 a.m. ten days ago from a noisy lay-by on the A3, and we are damn well going to finish it together. Having walked all this way together, around the biggest mountain in Western Europe, there simply is no other choice.

So it is agreed and having finished breakfast we prepare to leave the rather seductive comfort of Le Moulin and head out into the heaving rain. Having been spoiled by nine days of perfect weather this is all something of a shock to the system and it takes us a while to work out how best to prepare and to psych ourselves up for what is clearly going to be a very wet day on the trail. When walking the GR10 I swore by my trusty poncho in bad weather, a tough waterproof cape big enough to fit over both my rucksack and myself. The advantage of the poncho is that it allows a lot more freedom of movement than traditional waterproofs and is also more airy and, hence, less sweaty. Anyway, I have no choice as I left my raincoat in Rupert's car in Les Houches, judging it as expendable in my attempt to reduce the weight of my rucksack. I have also been waxing lyrical to Rupert about the merits of the said poncho, so he was minded to buy one for himself back in Courmayeur, based upon my recommendation. But come the time, as we stand gazing out of the door of the gîte at the rivers of water flowing down the road and the curtain of rain descending from the heavens, Rupert is not so sure. My conviction in the poncho is complete and so persuaded we help each other to pull the great billowing green cloaks over our heads and rucksacks, fighting with the poppers to get everything in place as it should be. Rupert's poncho seems to be a less robust version than mine and the poppers are not filling either of us with confidence, but we hope they will do. It is also too small, which just adds to the spectacle. Underneath my poncho, I am wearing my usual attire of shorts and T-shirt, my plan being to keep things nice and loose and airy as I have no doubt we will start to warm up as we get walking. Who knows, the sun might even come out.

All this poncho-related activity causes quite a flap in the gîte as the others also all prepare to head off. Ricardo is well sealed from the elements in his waterproof jacket and trousers, as is everyone else and it soon becomes clear that our choice of rainwear is the source of some amusement. As Rupert and I step out into the rain we look at best like a pair of dysfunctional superheroes and at worst, deranged hunchbacks. The daughter of our South African friend comes out in a fit of hysterics to take photos of us, saying that she wants to use them for a project she is doing at school – a PowerPoint presentation about cultural traits no less. We both have an uncomfortable vision of classes of children laughing uproariously at the crazy English. Shouldn't we be signing a consent form, I say to Rupert? Oh well.

We make our way up the road in the direction that we came from yesterday. Under normal conditions this stage of the trail does not look too arduous: a four-hour, eight-kilometre climb of around 730 metres up to the Refuge la Flégère. This is, in fact, the shortest stage of the TMB, but these are not normal conditions, compared to what we have become used to, and it is pouring down and really quite cold. Leaving the gîte and the little hamlet of Frasserands we walk back along the lane to Tré-le-Champ and then up a stony track to the main road that we walked down yesterday. It's strange to think that we could just hop on a bus and allow it to whisk us off down this road to Chamonix – a trip that would probably take no more than 20 minutes. I can feel Rupert twitching as we pass the bus stop. Crossing the road we come to a footpath signposted to Lac Blanc, La Flégère and Aiguillette d'Argentière and from here the path rises steadily up through woodland. The rain is bitingly cold and inside my poncho I feel somewhat disconnected and sealed off from the outside world, the hood magnifying the sounds of my breathing as if I were Darth Vader. Let's hope the force is with us today. I keep my head down, protecting my face from the wind and rain and push on, eager to keep moving. Rupert however is struggling with his flimsy poncho and has to repeatedly stop so he can rearrange and reattach the poppers that keep bursting open.

As I wait for him I feel the cold begin to insidiously wrap itself around me as the rain begins to soak my shorts that are exposed by the sharp, bitter wind. For the first half an hour we make spluttering progress until Rupert finally gives up on his poncho and changes al fresco into his waterproof jacket and trousers while I seek futile shelter under a tree, desperately trying to keep warm and dry in the absence of any other waterproofs. I can't remember it being like this in the Pyrenees and I begin to think that I may have underestimated the conditions today (or perhaps overestimated the power of my poncho), but reassure myself that once we get going we will soon warm up.

Pressing on, the path continues through the woodlands and then begins to open out to reveal craggy features of rock and stone rising up above us, the first indications of the Grand Balcon Sud that runs high above the Chamonix valley. The path takes us past the narrow, towering pillar of the Aiguillette d'Argentière, apparently a favourite with rock climbers, although we can see little in the deteriorating conditions and then we come to a single metal ladder, bolted into the rock, marking the beginning of the *passage délicat*. This section would have been great fun in any other conditions, but in the lashing wind and rain and fast reducing visibility it is hard to relish the experience. My hands are already cold and the metal handrails of the ladder are freezing to the touch as we make our way up the rock face one by one. It feels safe enough although ungainly as the weight of our rucksacks pulls us backwards and I am conscious that a simple slip or sudden loss of grip could be very nasty indeed. This section is made up of around seven or eight of these ladders, railings and wooden steps totalling a height of around 200 metres and I am sure the views, if we could see anything, would be stunning. As it is we focus on the task of reaching the top safely, Ricardo stopping to take a quick photo or two in the lashing rain.

At the top of the ladder section, the path rises again and there is a steepish climb to the Tête aux Vents, a junction in the path marked by a large boundary cairn. This is our high point for the day at 2,132 metres although there is an option to go further up to the Refuge du Lac Blanc, which on any other day is probably what we would have chosen to do. But up here at the Tête aux Vents the weather is deteriorating quickly and the temperature must have dropped a good 20 degrees Celsius from the temperature of the last few days; last night's storm is clearly heralding the arrival of a cold front. The rain starts to turn to sleet and then to hail, lashing relentlessly at our faces and a dense mist starts to close in around us. By now my poncho has been rendered pretty much useless, my clothes drenched and I compensate by walking more quickly to generate some body heat and cut down the time it will take us to reach La Flégère. Rupert and Ricardo are slower and each time I wait for them to catch up I feel my hands freezing and my body temperature seems to be plummeting rapidly.

The two of them want to stop for a while, to get their breath back and eat something while they hunker down behind a rock, but I need to keep moving so we agree that I will go on alone and meet them at La Flégère. I press on hoping that the trail will be easy enough to follow, particularly as the mist has closed in to a visibility of only about five metres. I can't feel my hands anymore and start worrying about frostbite so I alternate: one hand on my stick, the other in my mouth trying to breathe life into my numbed, lifeless fingers. My attempts to warm up are thwarted by my soaking clothes and I contemplate getting some dry clothes from my rucksack, although I am reluctant to stop and know that any dry clothes will be drenched in an instant. In the thickening mist I come to what looks like a junction in the path and become anxious about taking a wrong turning and I realise how foolish I have been in underestimating the change of weather, having been lulled into a false sense of security by the fantastic late summer weather of the last week. I hear the words of my GR10 friend Rob rattling around in my head: 'Never take risks with mountains' and wonder what he would say if he saw me up here in the French Alps at 2,132 metres in nothing but T-shirt and shorts at close to zero degrees with a useless poncho flapping around my ears. He would no doubt call me a nobber.

It doesn't take long to die of hypothermia. Depending upon air temperature and wind chill factor, the symptoms may not always be apparent at first, it can creep up on one insidiously. It begins with a natural feeling of cold accompanied by shivering – an instinctive reaction as the body tries to generate heat. A feeling of numbness then sets in as the shivering increases to a point where it soon becomes uncontrollable. Normal body temperature is 37 degrees Celsius, and when it drops to between 34–35 degrees Celsius one might experience mild confusion, apathy perhaps. Speech can become garbled and incoherent and the thought process begins to slow. As the body temperature drops further the shivering stops and muscular rigidity sets in, the body stiffening to the point where it may be hard to walk or stand, and mentally one may become increasingly incoherent, confused or irrational.

Approaching a body temperature of 32 degrees Celsius, uncovered skin can begin to swell and colour, feeling ice cold to the touch. At this stage a person might be semi-conscious, with dilated pupils and a barely registerable pulse. At 28 degrees

Celsius and below, unconsciousness sets in, the heart stops and that's your lot.

Now, I don't wish to be overdramatic, but as I stand freezing in the driving sleet at over 2,000 metres high on the Grand Balcon Sud of the Mont Blanc massif, in nothing but a T-shirt and shorts that are drenched to the core, wondering which path to take, I start to feel quite anxious. I'm shivering (stage one) and my hands and face feel numb (stage two?) and if I screw up here and take a wrong turn, with a visibility of less than five metres, it may not be too long before I hit stage three – and irrational thinking is the last thing I need. I am reminded momentarily of the Romantic poets and the notion of the 'sublime': the 'dark, confused, uncertain images' of Edmund Burke's imagination and the idea of the mountains being wonderfully wild and dangerous.

Casting such thoughts aside I fight the urge to just pick a path and keep on walking and instead decide to wait for Rupert and Ricardo, knowing that it should only be a short time before they come along. I listen out for their approaching voices, and when I eventually do hear voices they are not those of Rupert and Ricardo but of the young American couple who were also staying at Le Moulin and now appear like wraiths out of the mist. I explain that I am waiting for the others and they tell me that they are not far behind, but as the Americans head off I am caught by a moment's indecision and then quickly decide to follow them, not wanting to stand motionless in the cold for any longer than I need to. I keep the young couple ahead of me, making sure they stay within my range of visibility. They are walking fast, which suits me, and it is not long before we come to a stream that has been transformed into a flash flood of a torrent by the incessant rain. It's hard to cross and to slip and fall could lead to a nasty and very wet tumble downstream, and so we help each other across using the slippery point of one barely visible rock as a precarious stepping stone.

The next hour or so seems to last for an eternity as, keeping the Americans in view, I simply focus on my feet and the path, blocking out all other thoughts. At some point this will end, I tell myself, and it is just a case of tolerating the discomfort until it does. The fact that I have run many a ten-kilometre cross-country race over the years and the odd half marathon, often in atrocious winter conditions, probably helps. It is as much psychological as it is physical, about a mindset that, in the face of having to endure the intolerable, becomes almost dissociative, a split between mind and body.

Apparently, somewhere beyond the all-enveloping mist are some of the best views on the TMB. As Kev Reynolds helpfully points out, 'Nowhere along the Grand Balcon Sud does the panorama offer anything less than perfection.' Well, Kev, it's bloody well offering less than perfection now and even my feet, encased in my trusty Berghaus boots, are totally drenched from the rivers of mud and water that I am having to trail through. Eventually, I see signs of humanity, some kind of earthworks activity with great swathes of land dug away and piles of massive concrete pipes stacked to one side. I see tracks gouged out in the hillside beneath pylons and cables above and realise this must be part of the ski slopes, which means that the cable car can't be far away.

Then, like an inverted oasis (offering dryness instead of water), I see the beckoning lights of La Flégère cable car station itself with its attendant restaurant and associated facilities sitting just a short way above the refuge. It's a good sight, I can tell you. The Americans head straight down towards the refuge, but I head through the brightly lit glass doors of the station building from where I can keep an eye out for Rupert and Ricardo. Inside, I pull off my wet clothes and put on a dry shirt, fleece and pair of trousers and then warm my unfeeling hands under the dryer in the restroom. Ah, bliss.

About 15 minutes later the bedraggled, weather-worn shapes of Rupert and Ricardo emerge out of the mist like extras from *Dawn of the Dead* and I wave them into the warmth of the cable car station. The place seems pretty much deserted, but on inspection we discover that it is operational and that the next car down to Les Praz de Chamonix is leaving in about 20 minutes. Quickly taking stock, Rupert and I decide that we will indeed take the car down to the bottom while Ricardo – hard-core to the very end – decides to stay at the Refuge la Flégère and walk the final stage to Les Houches tomorrow and then meet us in the evening in Chamonix.

To be honest, I feel somewhat conflicted at this point, knowing that to take the cable car down the mountain means that we won't have entirely completed the loop around Mont Blanc, that we will fall short by about 20 kilometres. But we have to balance this with the fact that we are soaked through to the bone and have perhaps earned ourselves a few bonus points by walking in some pretty atrocious conditions when many would have called it a day. Also, Rupert walked today when he could have easily caught the bus back at Le Moulin and a deal is a deal – I want us to end this trip together. It's all about process remember; it's not necessarily where you get to but how you get there. The thing is the thing.

We walk down to the refuge, which sits just a couple of minutes below the cable car station and quickly eat a chunk of bread and cheese with Ricardo, slowly reviving ourselves and having a therapeutic debrief from this pretty hellish day on the trail. Ricardo says that he fell into the torrent of a stream that I had just about managed to negotiate with the Americans, but avoided being washed away due to his backside getting thankfully wedged between a couple of rocks. Rupert did well to abandon his poncho when he did and he will also do well to both never wear one again or listen to any advice I might give him. Oh well, we live and learn.

A creaking of metal and a hum of electricity signals the imminent arrival of the cable car and so we bid farewell to Ricardo and arrange to meet tomorrow in Chamonix, where we have pledged to find him a place to stay. Running back up the hill to the cable car station, we realise that we need to get a ticket so it is a mad dash to the little ticket office outside and then back inside to the cable car that is just about to make its return descent down the mountainside.

With a grind and a whir, the cable car slips free of its housing and swings out into the swirling rain and mist. We can't see a thing, but to be honest we don't care, we are just grateful to be whisked away from the cold and towards the promise of warmth, food and maybe even a hot shower. We both feel washed out – literally – and are shivering as pools of muddy water form around our feet on the floor of the car. It's funny, this is the second time I have escaped from the mountains in a cable car in these circumstances, the first being with Rob on the GR10 when we got caught out in a terrible storm and just managed to catch the last car down to Luchon, saving us a tortuous descent of several hours and about 1,200 metres. In a way it seems a somewhat inglorious end to our walk, but perhaps it is as good as it can be. We have been treated to ten days of spectacular weather, more than we could have hoped for, and on our final day the benign presence of Il Bianco, the White Lady, has given us a petulant, tempestuous flash of her darker, more temperamental side; a little taste of her potential to switch mood in an instant. It is as if she has given us a kick up the arse to send us on our way and to remind us who's in charge up here in the high grounds, that this is her domain and let us not forget it. It's been a sobering day, that's for sure and I have been foolish in my underestimation of the conditions, so I hope I have learned a lesson for another day.

The cable car deposits us in Les Praz, a small village in the middle of the Chamonix valley and part of the commune of Chamonix. From here we take a bus into Chamonix itself where we plan to find somewhere to stay, but in our dampened, post-walk stupor we miss our stop and sail right through town. Re-evaluating our plan, we decide to go on to Les Houches where we can stay tonight and pick up Rupert's car tomorrow and drive back to Chamonix. This involves another bus and much waiting around, but eventually we find ourselves in Les Houches, full circle from where we began ten days ago.

I experience a strange sense of ambivalence, a feeling of triumph tinged with sadness that our adventure has come to an end and that our little TMB bubble is about to burst, although I am very much looking forward to seeing Nicky and Jess again. But like our cold, wet, numbed bodies it is hard to feel too much right now beyond the immediate need for comfort. We both want to get dry and warm, find something to eat and then we can allow ourselves to reflect upon our achievement. Getting off the bus and walking down through the village we call in at the tourist office to ask about a cheap place to stay and the friendly woman tells us about the nearby Gîte Michel Fagot, and as she explains where it is a charismatic, amusing character blows in through the door. It none other than M. Michel Fagot himself. The tourist office woman, doing a great networking job, introduces us to each other and we explain to M. Fagot that we have had a long, wet day on the trail and are in need of warmth and shelter. 'No problem,' he says with a grin, 'unlimited hot water,' and points us in the direction of his establishment.

The gîte is great with its welcoming blue shutters and overall homely sense of character. At €17 each it is not only the cheapest place we have stayed in, but also one of the best, ironic considering we are back in Les Houches where we started. Like most gîtes we have visited, the key is a mix of the functional and the comfortable. The exception is the Elisabetta, which was neither functional nor comfortable but an experience all the same. Inside, the place seems more or less empty; a large, informal dining area adjoins a kitchen where we can cook if we choose. Upstairs we have a small dormitory all to ourselves, and downstairs in the basement is a washing and drying room where we can hang all our wet clothes. All in all, it's exactly what we need and after a bit of faffing about I indulge in one of the longest, hottest, most pleasurable showers I have ever experienced. It feels a long way from nearly freezing to death up on the Grand Balcon Sud.

Later, feeling revived and recuperated, we go for a couple of beers in a nearby bar and then buy some food from the supermarket to cook back at the gîte. And so later, after a few more beers and a bottle of red wine and a tasty home-cooked meal of pasta in tomato sauce, we allow ourselves to drift into a post TMB fuggy haze of congratulation and toast ourselves, several times, to the fact that we have (just about) walked around the highest mountain in Western Europe. It has been a privilege to walk with Rupert, he is a great friend and companion. It can be something of a test to spend long days walking with someone; you never quite know how you may get on, but I have enjoyed his company and I think perhaps he has taught me something about how to slow down and enjoy the moment; that it is OK to stop and sketch, so to speak, rather than press on relentlessly. I am innately restless, for better or worse, so I need to accept this lesson.

Replete with wine and food we chat for a while with our host Michel and he introduces us to Radio FIP, a French station that plays a fantastically eclectic range of non-stop music, especially jazz, and so we are treated to a fine dose of Miles Davis' seminal album *Kind of Blue* to end the evening and indeed our walk. If anyone was to embody our philosophy of the last ten days – the thing is the thing – it would be Miles Davis with his laid-back, spacious, freewheeling trumpet that fills the room with its rich, sonorous tone.

Twenty-two

The next morning, after grabbing a quick coffee and croissant in a nearby cafe, we pick up Rupert's car (which is thankfully still safely parked where we left it) and drive the short distance to Chamonix, where we find a budget hotel on the outside of town. We book Ricardo and ourselves in, and then walk into town. Later, while Rupert goes off to sketch the impressive church that sits proudly in the centre of town next to the tourist office and to wait for Lucy who is on her way over on a bus from Geneva, I grab myself a beer and a croque-monsieur and hang out for a while in a cafe. It's pleasant, sitting outside as the early afternoon sun battles with the clouds that are forming like candyfloss around the peaks of the mountains, which glisten brightly after their overnight dusting of fresh snow. Chamonix is a bustling, classy place and the crowds quickly start to pick up as the town becomes a veritable hive of multinational activity, and I catch conversational snatches of French, Italian, German and American among many others as people promenade past the cafe.

With the ever present backdrop of Mont Blanc, Chamonix is essentially the home of mountaineering and the major tourist centre of the Alps, busy in both summer and winter as it widens its appeal to embrace all manner of mountain-related activity. In the middle of town stands an impressive statue of our friends Horace-Bénédict de Saussure and Jacques Balmat, the latter raising an arm and pointing Saussure towards the summit of Mont Blanc, and housed in a former palace, the Musée Alpin retraces the history of the town from when the first 'tourists' came to marvel at the spectacle of the glaciers back in the eighteenth century.

Chamonix was also the home of the first Winter Olympics in 1924 and in the closing ceremony of the games Pierre de Coubertin, the founder of the Olympic Movement and the International Olympic Committee, presented an honorary 'prize for alpinism' to Charles Granville Bruce who led the second attempt on Everest in 1922. The attempt was unsuccessful, but did establish a new world record climbing height of 8,230 metres. Tragically, seven porters lost their lives in an avalanche during this endeavour; the first reported climbing deaths on Everest. The award was a nice nod of recognition to the growing alpinist community of the time and perhaps an acknowledgement of the people who lost their lives on the expedition.

Chamonix is well known for its spectacular cable car that takes you the 3,842 metres to the Aiguille du Midi. Back in 1955 when it was first constructed, it was the highest cable car in the world and together with the cable car system that carries you to Pointe Helbronner from Entréves in the Aosta valley, it is actually possible in the summer to cross the entire Mont Blanc massif by cable car.

As an occasional reminder of yesterday's weather, the wind picks up and a sudden gust sweeps the contents of several tables to the ground and although the sun is shining there is quite a chill in the air. As more people pour into town I see a number of our fellow TMB comrades who have begun to drop down from the hills above, faces that I recognise but which are now somehow out of context without the hiking accoutrements of rucksacks and walking poles. The lovely South African family pass by, their children delightful, and we chat for a while about our respective experiences on the trail. I ask the children how they are getting on with their school project and they rather unsettlingly reassure me (if that's not a contradiction in terms) that Rupert and I will be the main focus of their presentation. I ask their father, a fellow lecturer, if his university's ethical committee has signed off his children's project, but he just shrugs this off. They, of course, don't know that they will feature in this book on the TMB, so I guess that makes us quits.

A little later I see the Belgian South African, with his grizzled features and impressive white moustache, who I last crossed paths with back in Courmayeur. Claire and Julie stroll by, joined by Claire's boyfriend, and soon followed by Ricardo and Rupert, who has met up with his partner Lucy. It's good to be all together again and as we chat to all these people who we have walked with off and on over the course of the last ten days, it seems that many of them have stories about Rupert and myself – indeed it seems that we have acquired a certain degree of eccentrically tinged infamy during our course around Mont Blanc. Me with my floral trilby and wooden stick, and Rupert with his sketch pad and stories of ponchos and the Rifugio Elisabetta. Oh well, we are happy to have provided some light relief along the way.

Later, Rupert, Lucy and I stroll down to the restaurant where several tables have been taken up by various members of the TMB collective, who have by both design and accident, gathered this evening to celebrate our shared achievement. Claire and Julie are in residence, our RSPB friends, and it is great to meet up with them again, joined by Claire's boyfriend – a lovely unassuming Scottish chap. We really enjoyed their company during the days that we walked together, with their vibrant personalities and quirky sense of humour and, of course, it means that our little marmot gang is temporarily reunited. Ricardo is here, TMB networker extraordinaire and someone who has become a good friend on the trail with his impish charm and colourful stories. The young American couple are here too, with whom I shared an uncomfortable hour or so on the Grand Balcon Sud yesterday, which we acknowledge with a brief nod and a smile – no words necessary. And on the table next to us are the northern group that we dovetailed with for a while on the stage from La Fouly, full of friendly cheer and Yorkshire bonhomie.

There is a strong sense of camaraderie between us all, a sense of feeling joined together – bonded – by our mutual experience on the TMB. We have all been brought together here in this little restaurant in Chamonix, which sits at the foot of Mont Blanc – the summit lost somewhere in the swirling darkness above – by our love of walking. It really is the art of simplicity; a long stroll around a big mountain, but the path also connects us together, albeit momentarily, and it is a moment to savour – another juncture among the myriad pathways that illuminate our landscape like little glowing heat trails of humanity. Personally, this has been a journey that has taken me both outside and within, something of a retrospective as with each step forward my mind has taken me back another two. Maybe it is because my fiftieth year looms large, akin to a mountain pass that marks a border into new unchartered territory and signposted by a cairn of old weathered stones picked up along the way. In this sense it really has been a circuitous journey, taking me full circle in more ways than one. 'Vague memories hang about the mind like cobwebs,' wrote George Eliot in *Romola*, and indeed I feel that I have managed to blow away some of those cobwebs along the way; cleared out some of those dusty corners. Most of all though, it is Mont Blanc itself who has brought us all here, this great timeless mountain that sits proudly in the Western Alps, and while us mere mortals have drifted in and out of each other's company during our circumnavigation she has maintained her regal presence.

As we sit together contently chatting and swapping stories about experiences on the trail, the waiter comes round and hands out menus, which we study with a hungry interest. Someone asks Ricardo what he is thinking of ordering and he theatrically clears his throat, effectively silencing the room. 'Well,' he says, studiously examining the menu, 'I think I will have the Chicken Tarka.' He leaves sufficient space, waiting for his cue while everyone looks at him with puzzled curiosity. 'Don't you mean Chicken Tikka?' I say innocently. 'No,' he says with a perfectly executed dramatic pause. 'It's like a Chicken Tikka, but just a little otter.'

Notes

1 Reynolds, Kev *Tour of Mont Blanc: Complete Two-way Trekking Guide* (2011, reprinted 2014, Cicerone).

2 Freshfield, Douglas William with the collaboration of Henry F. Montagnier *The Life of Horace Benedict De Saussure*. (2012, HardPress Publishing facsimile print of the University of Toronto Library book published in London by Edward Arnold in 1920).

3 Saussure, H *Voyages Dans Les Alpes*. Barde, Manget, Geneva, 1786 Vol IV page 219 taken from Fleming, Fergus *Killing Dragons: The Conquest of the Alps* (2000, Granta).

4 Fleming, Fergus *Killing Dragons: The Conquest of the Alps* (2000, Granta).

5 Paccard, Jacques – his account of the 1783 ascent taken from Fleming, Fergus *Killing Dragons: The Conquest of the Alps* (2000, Granta). A copy of Paccard's journal is held at The Alpine Club, London.

6 Milner, C. Douglas *Mont Blanc and the Aiguilles* (1955, Robert Hale First Edition).

7 Payot, P. *Au Royaume du Mont Blanc* (1996, La Fontaine de Siloe, Montmelian) page 28 taken from Fleming, Fergus *Killing Dragons: The Conquest of the Alps* (2000, Granta).

8 Mathews, C. E. *The Annals of Mont Blanc* (1898, T. Fisher Unwin).

9 Rousseau, Jean-Jacques; trans. McDowell, Judith H. *La Nouvelle Héloïse: Julie, or the New Eloise: Letters of Two Lovers, Inhabitants of a Small Town at the Foot of the Alps* (abridged edition 1990, Penn State University Press).

10 Rousseau, Jean-Jacques; trans. Cohen, J. M. *The Confessions* (reprint edition 1953, Penguin Classics).

11 Macfarlane, Robert *Mountains of the Mind: A History of a Fascination* (2008, Granta) page 163.

12 Harvey, William *The Peasants of Chamouni: Containing an Attempt to Reach the Summit of Mont Blanc, and a Delineation of the Scenery Among the Alps* 1826 (2010, Kessinger Publishing).

13 Beattie, Andrew *The Alps: A Cultural History (Landscapes of the Imagination)* (2006, Signal Books) page 208.

14 Bonatti, Walter *The Mountains of My Life* (2010, Penguin Classics).

15 Csikszentmihalyi, Mihaly *Flow: The Psychology of Optimal Experience* (2008, Harper Perennial Modern Classics).

16 Nietzche, Freidrich taken from Fleming, Fergus *Killing Dragons: The Conquest of the Alps* (2000, Granta) page 91.

17 Macfarlane, Robert *The Old Ways: A Journey on Foot* (2012, Hamish Hamilton).

18 Saussure, H *Voyages Dans Les Alpes*. Barde, Manget, Geneva, 1786 Vol IV page 175 taken from page 57 Fleming, Fergus *Killing Dragons: The Conquest of the Alps* (2000, Granta)

19 Smith, Albert – his account of his ascent of Mont Blanc taken from Fleming, Fergus *Killing Dragons: The Conquest of the Alps* (2000, Granta) pages 154–155.

20 Whymper, Edward *Scrambles Amongst the Alps in the Years 1860–69* (2002, National Geographic Books).

Further reading

Bryson, Bill *A Walk in the Woods* (1998, Black Swan).

Burke, Edmund *A Philosophical Enquiry into the Origin of Our Ideas of the Sublime and Beautiful* (2008 Oxford Paperbacks)

Von Daniken, Eric *Chariots of the Gods* (1990 Souvenir Press)

And also many thanks to the Alpine Club, Charlotte Road, London

THE HAIRY HIKERS

A Coast-to-Coast Trek Along the French Pyrenees

David Le Vay

£8.99

Paperback

ISBN: 978-1-84024-237-2

With a glint in his eye, Rob turns and asks me if I want to 'touch his furry puma'. We are only hours into the trip and things have already taken a sinister turn. Thankfully, it turns out he is referring to the little embossed logo on his new shirt.

Fuelled by a degree of midlife crisis and the need to escape from routine, armed with rusty schoolboy French and plenty of schoolboy humour, friends David and Rob set out to walk the fabled GR10 hiking trail. It will take them from Hendaye on the Atlantic coast to Banyuls-sur-Mer on the Mediterranean, through beautiful scenery and one of the most spectacular mountain ranges in Europe. Just about perfect – if you can put aside the inevitable snoring-induced conflict and bad habits that result from two men spending over seven weeks in each other's company.

A BEST-SELLER IN GERMANY
A
YEAR
IN THE
SCHEISSE
Getting to Know the
Germans
ROGER BOYES

A YEAR IN THE SCHEISSE

Getting to Know the Germans

Roger Boyes

£9.99

Paperback

ISBN: 978-1-84024-648-3

This is the story of an English journalist's absurd adventures living in Germany. Facing bankruptcy, Roger is advised by his accountant to make use of a legal loophole: in Germany married couples have their tax bill halved. So the search is on for a bride. Meanwhile his father, a former war hero, is also in financial trouble and is threatening to move to Germany and sponge off his son. The combination of crises sets in motion a hilarious romp during which we discover more than we really wanted to about German nudist beaches, the British media's obsession with Adolf Hitler and how to cheat at the Berlin marathon.

'amusing story of an English journalist's adventures... will leave you with a smile on your face!' *Kirriemuir Herald*, June 2008

'I scheissed myself laughing. Herr Boyes has written a thigh-slapper of a book' Henning Wehn

TALES FROM THE

FAST TRAINS

EUROPE AT 186 MPH

TOM CHESSHYRE

TALES FROM THE FAST TRAINS

Europe at 186 mph

Tom Chesshyre

£8.99

Paperback

ISBN: 978-1-84953-151-1

Tired of airport security queues, delays and all those extra taxes and charges, Tom Chesshyre embarks on a series of high-speed adventures across the Continent on its fast trains instead. He discovers the hidden delights of mysterious Luxembourg, super-trendy Rotterdam, much-maligned Frankfurt and lovely lakeside Lausanne, via a pop concert in Lille.

It's 186 mph all the way – well, apart from a power cut in the Channel Tunnel on the way to Antwerp. What fun can you have at the ends of the lines? Jump on board and find out...

'Make a note to buy this... If you're tired of the endless delays and extra costs at airports, this book will inspire you to hop on a train. Discover hidden delights of Europe with no hidden charges or taxes. It's a fun-packed read, as well as being very informative.' *Prima*

'If you've "done" Paris and Bruges and are wondering, "Where next?", then this may be a quiet revelation' Andrew Marr

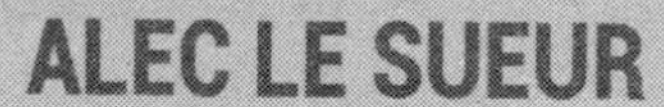

author of THE HOTEL ON THE ROOF OF THE WORLD

BOTTOMS UP IN BELGIUM

Seeking the High Points of the Low Lands

BOTTOMS UP IN BELGIUM

Seeking the High Points of the Low Lands

Alec Le Sueur

£8.99

Paperback

ISBN: 978-1-84953-247-1

Brussels and all those Eurocrats on the gravy train? It's just so boring. Why, you can't even name ten famous Belgians!

Until 1993, Alec had never been to Belgium, so it came as some surprise when in August that year he found himself at the altar of a small church in Flanders, reciting wedding vows in Flemish. It was the start, for better or for worse, of a long relationship with this unassuming and much maligned little country. He decided to put worldwide opinion to the test: is Belgium really as boring as people say it is?

Immersing himself in Belgian culture – and sampling the local beer and 'cat poo' coffee along the way – he discovers a country of contradictions, of Michelin stars and mechanically recovered meat, where Trappist monks make the best beer in the world and grown men partake in vertical archery and watch cockerels sing (not necessarily at the same time).

This colourful and eccentric jaunt is proof that Belgium isn't just a load of waffle.